LIVING with WILDLIFE

This book is dedicated to our Australian wildlife, navigating the challenges of our shared landscape; and to all those people working to care for them.

I would like to acknowledge and extend my appreciation for the Dja Dja Wurrung People, the Traditional Owners of the land upon which this work was written. I would like to pay my respects to leaders and Elders past, present and emerging for they hold the memories, the traditions, the culture and the hopes of all Dja Dja Wurrung People. I express gratitude in the sharing of this land, sorrow for the personal, spiritual and cultural costs of that sharing and hope that we may walk forward together in harmony and in the spirit of healing.

LIVING with WILDLIFE

A GUIDE FOR OUR HOMES AND BACKYARDS

TANYA LOOS

PUBLISHING

A catalogue record for this book is available from the National Library of Australia

ISBN: 9781486316946 (pbk)
ISBN: 9781486316953 (epdf)
ISBN: 9781486316960 (epub)

Published by:

CSIRO Publishing
36 Gardiner Road, Clayton VIC 3168
Private Bag 10, Clayton South VIC 3169
Australia

Telephone: +61 3 9545 8400
Email: publishing.sales@csiro.au
Website: www.publish.csiro.au
Sign up to our email alerts: publish.csiro.au/earlyalert

Front cover: Green tree frog (photo by Ethan Mann)
Back cover: (top to bottom) Red-necked wallaby at bird bath (photo by Sandid/Pixabay), Ringtail possum (photo by Ethan Mann), Cassowary on verandah (photo by Patrick Tomkins)

House icon on p. 1 by Sinhara N S/Shutterstock.com

Edited by Joy Window
Cover and text design by Cath Pirret
Typeset by Envisage Information Technology
Index by Indexicana
Printed in China by 1010 Printing International Ltd

CSIRO acknowledges the Traditional Owners of the lands that we live and work on across Australia and pays its respect to Elders past and present. CSIRO recognises that Aboriginal and Torres Strait Islander peoples have made and will continue to make extraordinary contributions to all aspects of Australian life including culture, economy and science. CSIRO is committed to reconciliation and demonstrating respect for Indigenous knowledge and science. The use of Western science in this publication should not be interpreted as diminishing the knowledge of plants, animals and environment from Indigenous ecological knowledge systems.

The paper this book is printed on is in accordance with the standards of the Forest Stewardship Council® and other controlled material. The FSC® promotes environmentally responsible, socially beneficial and economically viable management of the world's forests.

Jul24_01

Contents

Acknowledgements

First of all, a huge thank you to Briana Melideo and Mark Hamilton at CSIRO Publishing – Briana for helping to craft and shape the project, and Mark for his gentle shepherding and patience. Thanks also to copy editor Joy Window, and to the rest of the team who have been such a pleasure to work with.

I would like to thank Ian Temby for his support throughout. Ian assisted with general discussions and fact checking and editing for many of the entries. Dr Pia Lentini and Dr Kylie Soanes's work on nature in cities is a total inspiration and the basis for much of the work in this book. Doug Gimesy, Millie Ross and Louise Nicholas assisted with early discussions and advice. Support from comms legend Dominica Mack, IFAW, early in the project meant the world to me. Thank you all so much!

I would also like to give a shout out to Andrew Masterson, former editor of *Cosmos* magazine. A 2-year stint writing for Andrew in print and online gave my work an immense boost, and led to this dream-come-true project.

The contributions of the following people were invaluable and I am very grateful for their generous time and expertise, and in some cases images:

- **Wildlife rehabbers and hospital staff:** Greg Irons, Bonorong Wildlife Sanctuary. Jenny Mclean, Tolga Bat Rescue and Research. Carol Jackson and Tasha Rea, Kanyana Wildlife Rehabilitation Centre. Dr Amber Gillett, Australia Zoo Wildlife Hospital. Dean Huxley, WA Wildlife. Nalini Scarfe, Boobook Wildlife Shelter. The staff at Wildlife Health Australia who produce their wonderful fact sheets.
- **Conservationists:** Janine Duffy, Koala Clancy Foundation. Rebecca Keeble, IFAW. Peter Rowles, C4: Community for Cassowary and Coastal Conservation. Alan Henderson, Minibeast Wildlife. Tori Seydel and Brooklyn Dysart, Wildlife Watcher, an initiative of Social Marketing @ Griffith.
- **Ecologists and biologists:** Dr Ana Gracanin, Justin A. Welbergen, Dr Johanne Martens, Simon Watharow, Nadiah Roslan, Dr Jodie Rowley, Dr Holly Parsons, Dr Bronte Van Helden, Ash Miller, Robin Rowland, Dr Chantelle Derez, Dr Lucy Aplin, Prof Dr Barbara Klump, and finally Alexandra Holm and her team of admins on the Australian Snake Identification, Education and Advocacy (ASIEA) Facebook page.

- **Communications staff:** Natalie Filmer, Zoos Victoria, Aleisha Hall, Wildlife Victoria, John Grant, WIRES and Dhwani Chandra from Australia Zoo.
- Kiara Mucci for her wonderful line drawings. Patrick Tomkins, Saskia Granger and Ethan Mann for their generosity in sharing their incredible photos. Bonorong Wildlife Sanctuary, Australia Zoo, Wildlife Warriors, Wildlife Victoria, Doug Gimesy, Bill Sheaffe, Helen Greenwood, Donna Pomeroy and Niel Wark were also very kind to supply their images. Thank you!

Thank you also to my friends near and far, and mum Evelyn and sister Jacinta for their love and encouragement. And last but not least, to my darling husband Chris for his support throughout the project, including putting in long hours helping me finalise the manuscript once we got to the pointy end.

Introduction

I was one of those lucky kids who fell in love with nature and wildlife at a young age. I was born in Calgary, Canada, and would go on to move and live around the world as my dad worked in the oil industry, in mining and exploration. After a short stay in Libya, in 1980 we settled in London for a year. David Attenborough's incredible *Life on Earth* was on the telly, and a friend of my mum's gave me a book on pond life. I was 6 years old and completely smitten with the natural world.

By the time I was 10, I had binoculars and a huge library of nature and wildlife books. I was also an avid nature journal writer and avid reader, trained in how to be an amateur naturalist by Gerald Durrell, and in ecology and evolution by David Attenborough.

When I was 9–12 years old, we lived in Jakarta, Indonesia, and the 'summer' holidays of my school fell in the Australian winter, which I spent in the seaside town of Caloundra, Queensland. Each day I would wobble off down the road on my bike with

A critically endangered western ringtail possum at home in their habitat – in this case above the shower block of a busy caravan park in Busselton.

ETHAN MANN

The author as a young naturalist, Caloundra 1986.

MARCEL LOOS

my binoculars, notebook and packed lunch, and spend glorious hours birdwatching and nature spotting in extensive paperbark swamps, mangroves and mud flats.

My favourite place was an estuary, a mangrove-lined riverbank away from the caravan parks and fishing boats by the shore filled with the peaceful sights and sounds of birds fishing for their food: herons, kites, waders, kingfishers. One day, at the age of 11, I cycled to this spot and the entire river's edge on both sides had been bulldozed and was devoid of all vegetation. Only my favourite wattle, an excellent perching spot for birdwatching, remained. I sat in it and cried.

I knew the clearing was to build houses. I knew that I lived in a house, and that all the streets and houses of Caloundra had also begun with the same destruction. And it just didn't seem right, or fair, but it seemed inevitable.

As I have grown older, I have come to realise that 'humans equals destruction and death for nature' is not a grim inevitability. There is, in fact, so much we can do.

There are so many choices that we can make to reduce harm, and truly coexist, survive and thrive with our wonderful wildlife.

A search online for advice on wildlife at home often comes under the heading 'living with wildlife'. This approach encourages landholders to understand and care about their local wildlife and solve any problems we may have with our wild neighbours without resorting to lethal control methods, and ideally with the least amount of harm to the animal. *Living with Wildlife* helps us live more harmoniously with our neighbours.

This book takes this approach a step further with a focus on wildlife-friendly actions that celebrate our homes as a part of their habitat.

An urbanised Australia – we live where wildlife lives

The island we call Australia was once a vast tropical rainforest, with three species of koala, giant wombats, platypuses, lizards and even flamingos in our inland seas. By the time we reached the Holocene some 11,000 years ago, these rainforests had retreated to tiny pockets on the east coast, and the woodlands, wetlands and forests clung to the moister outer edges of the continent.

Following the European invasion of Australia, the creation of agricultural and urban areas occurred on the richest, most productive land. Melbourne, for example, was once a vibrant confluence of the Yarra and Maribyrnong rivers, clothed in wetlands and woodlands rather than roads and buildings.

Today, Australia is one of the most highly urbanised countries in the world: 85% of us live in townships of 10,000 people or more.[1] And, rather than occupying a vacuum, our human settlements continue to take up land that is prime wildlife habitat – as I witnessed that fateful day back in 1986. And I am sure many of you have also witnessed such destruction. The productive forests of coastal South East Queensland are rich in habitat resources for koalas and grey-headed flying-foxes – and today these forests are also the site of a growing mega-city that stretches from the Gold Coast to Brisbane to the Sunshine Coast. So, often when we have wildlife encounters in urban areas, it's important to realise they are not in our homes – we are in theirs!

Wildlife under pressure

Since the Black Summer megafires of 2019–20 and the subsequent floods, the repeated use of the word 'unprecedented' in media reports has come to signify the current era of climate change or climate breakdown, where each year we see extreme weather events such as landscape-scale bushfires, heat waves, flooding and storms. These events kill animals outright and reduce habitat for the wildlife that survives.

Adding to the above climate stressors, the human population is rising, especially on the east and south-west coastal areas, creating increased urbanisation and densification. While the images of flood or fire-stricken animals garner plenty of media attention, and quite rightly tug at our heart strings, human-induced or anthropogenic pressures facing wildlife every single day cumulatively outweigh the effects of extreme weather events.

Several multi-year studies in Victoria, Queensland, New South Wales and Tasmania have analysed the reasons that wildlife may require rescue in these states. The top reasons for animals coming into care are car strike (being hit by a car), becoming orphaned, dog or cat attack, entanglement in fence or netting, and disease. Of these survivors, across the four states, over 50% of the animals admitted into care require euthanasia, as their injuries or disease are too great. These studies also only count the *survivors* of the trauma, as an animal killed by hitting a window or being attacked by a cat is already dead, and therefore does not need rescuing.[2,3,4,5]

All of these anthropogenic threats increase as human population increases, as urban sprawl encroaches into wildlife habitat and as the effects of climate change intensify. The *Australia State of the Environment Report 2021* painted a grim picture of the state of biodiversity and wildlife conservation in Australia with rising numbers of threatened species and several ecosystems on the brink of collapse. But rather than the environment being something we need to save 'out there', the report highlighted the fact that we can – even those of us in urban areas – play an important role by caring for the flora and fauna that live where we live. In fact, Australia's cities and towns are home to more than 96% of our population and 46% of threatened species.[6,7] While threatened species habitat includes the pockets of natural ecosystems or remnants clinging on in tiny reserves or the edge of land such as railways, our backyards also support threatened species such as grey-headed flying-foxes. And urban habitats such as parks and gardens can act as refuges during and after extreme weather events.

Living with wildlife in a changing world

The study of the troubled interactions between humans and wildlife is known as the study of human–wildlife conflict. The last 10 years or so have seen a shift from a focus on resolving conflict between humans and wildlife towards a new model of coexistence. Coexistence is defined as a 'dynamic but sustainable state in which humans and [wildlife] adapt to living in shared landscapes'.[8] Examples of this include farmers in the United States who have learned to ranch sheep without killing wolves, people in India who are using beehives to keep hungry elephants from their crops and Canadians using bear-proof containers to keep conflict between campers and bears at a minimum.[9]

The wildlife-friendly house and backyard is an active acknowledgement that our cities, towns, suburbs and bush blocks are shared landscapes. And while we don't have wolves, elephants or bears to contend with, the Australian context provides so much opportunity to explore what coexistence really looks like when we share our backyards and sometimes even our houses with rosellas, quolls, bandicoots and bats.

As we have discussed, the wildlife that lives where we live are facing ongoing habitat destruction, human-caused threats such as dog and cat attack, entanglement, car strike and increasingly extreme and frequent weather events such as heatwaves, floods and bushfires. Wallabies, snakes, possums and parrots do not make the distinction between natural and human-created habitats. While a visiting koala is welcome, but a brown snake is less so, both are in our backyards because our homes are encroaching into their habitat. Other species, such as brushtail possums and sulphur-crested cockatoos can delight and frustrate us in equal measure as they have learned to take advantage of the food and shelter our houses and gardens provide.

As we modify the land with our cities and towns, we create new habitats that some species have been able to exploit. Unconventional habitats are defined as areas

It takes a village! Ember the koala was badly burned in the 2019 bushfires and nursed back to health by Currumbin Wildlife Hospital and Friends of the Koala, supported by IFAW Australia. She was returned to the wild after months of intensive rehabilitation for burns, smoke inhalation and starvation. Ember was spotted with a healthy young joey in September 2021, but unfortunately, despite her recovery from the bushfires, in early 2024 she fell ill and passed away.

BILL SHEAFFE

originally created for human use that provide important habitat or resources for native biodiversity.[10] These unconventional habitats are not as well studied as large natural areas, but increasingly their importance for wildlife and particularly endangered species is being realised.

Research on the critically endangered western ringtail in Western Australia showed that Albany backyards are vital habitat for this species – even backyards with exotic rather than native vegetation! The study also discovered that, unlike previously thought, backyards far from natural area reserves could support these small possums. This means that anyone with a backyard in Albany is on the frontline of ensuring the ongoing survival of western ringtail possums.[11]

Other threatened species in urban areas now partly dependent on unconventional habitats include the grey-headed flying-fox and the swift parrot. These species are nectar specialists and feed on the mass flowering of planted eucalypts of our cities' streets and car parks and may both be found in backyards.

Our unconventional garden habitats also help wildlife in hard times. During the landscape-scale bushfires and heat waves, our gardens are a vital lifeline for species such as gang-gang cockatoos, koalas and superb lyrebirds, providing shade, shelter, water and food resources.

In short, a wildlife-friendly house and backyard creates a safe haven for our wildlife as we navigate these next few decades together in a time of climate breakdown.

About this book

The major threats facing Australia wildlife such as climate change and habitat destruction require swift and well-funded action across all levels of government, business and the community. To most nature and wildlife lovers, tackling these threats can seem overwhelming: but this book is designed to offer hope! There is plenty of 'low-hanging fruit' – actions we can take around the house and garden that foster coexistence between humans and wildlife to the benefit of both parties.

While researching this book, I had the great fortune to interview some of the leading lights of wildlife rescue and rehabilitation around Australia. When I spoke with Greg Irons from the Bonorong Wildlife Sanctuary in Tasmania, Greg talked about Bonorong's successful rescuer training program, which has trained some 22,000 locals and visitors to Tasmania. Greg said that while it's nice to help, and people feel warm and fuzzy being rescuers, it is also equally important to ask how can we be 'preventers'? What actions can we take to *prevent* wildlife from getting into trouble in the first place?

Living with Wildlife builds upon the wealth of information out there on gardens for wildlife by putting the emphasis on the creation of a safe haven for your local wildlife, while also acting as a handy guide to solving wildlife 'problems' if your wonderful wildlife park is causing a bit of strife – either for you or for the animal. Ultimately, a wildlife-friendly house and backyard can help us *prevent* wildlife from succumbing to injury and disease. We can work together to reduce suffering and reduce the wildlife death toll!

In wildlife rescue, anyone outside the wildlife rescue sector reporting an animal is known as a 'member of the public' (MOP). While a dry term, it is very apt in that it tells us that everyone has a role to play. Taking an animal in trouble to a vet or wildlife carer may be the closest a person and their family has been to a wild animal and presents a beautiful opportunity for increased connection to all animals and their habitats. A MOP can be a landholder, a renter, an employee at a school, or local council, a tourist, a motorist – all of these roles and more can and do involve interactions with wildlife on an everyday basis. This book is written then for the MOP, and for ease of use will be referred to as 'you' throughout the text. For ease of language this book uses they/them pronouns when talking about an animal, unless the sex of the animal is known.

The book is divided into three sections: 'house', which covers commonly asked questions related to animals living in or visiting your home such as 'what do I do if there are bats or possums in the roof?' The 'backyard' section features what to do when you have garden visitors, from the friendly (e.g. echidna) to not-so-friendly (e.g. cassowary)! The backyard section also assists with problem solving regarding some of our more destructive wild friends such as cheeky sulphur-crested cockatoos. The final section relates to helping wildlife in trouble, and features practical advice on wildlife rescue, both every day and during extreme weather events. Topics such as wildlife friendly pet ownership, fencing, driving and why feeding wildlife is not recommended are also covered.

The wildlife-friendly solutions and living with wildlife advice in this book is designed to:

- fact check and standardise the advice related to commonly asked questions about the wildlife we encounter in our homes, backyards and neighbourhoods. References may be found at the end of chapters, with a 'Further reading' list at the back of the book.
- encourage first a 'just embrace it' approach (as long as it is safe for both animal and human!). For example, frogs in the toilet cistern are seasonal and will simply go back from whence they came (see pp. 9–15).
- encourage good design in the form of wildlife-friendly barriers. These are always the most humane solution – ensure the quoll cannot access your chickens (pp. 96–102), or the parrots and flying-foxes cannot get your fruit (pp. 69–76).

- teach the observation and problem-solving skills necessary for a wildlife-friendly approach: what is the animal doing (and why are they doing it)? How are they using the space? Has their behaviour changed? Can the garden or house be modified to solve the problem? Or what behaviour change do *you* as a human and manager of the space need to do?
- solve the problem on your property – solutions are best done within the animal's territory. Live capture and release is not good for possums (pp. 32–37) or snakes (pp. 111–116).

Lethal control, and deterrents such as bird spikes, ultrasonic deterrents, bird 'frite' cartridges and chemical deterrents, are not part of the wildlife-friendly approach – and will not be covered in this book. The land that we call 'Australia' is very big indeed, and while this book seeks to cover wildlife questions and encounters across a broad suite of habitats, it is no substitute for the wealth of knowledge held by the good people of your local wildlife rescue organisation or wildlife hospital. The staff and volunteers will have a keen understanding of local issues, and how to tackle them – so use this book as a prompt to get in touch with them.

Note that, just as our wildlife is under enormous stress from the rising pressures described in this chapter, so too are our vets, vet nurses, rescuers and rehabilitators. Patience, kindness and understanding will go a long way when seeking help, and there are many ways that you can support your local wildlife rescue organisation or wildlife hospital, so visit their website or reach out to them to see how you can assist.

References

1. Commonwealth of Australia (2019) 'Planning for Australia's Future Population'. Centre for Population, Canberra, Australia.
2. Heathcote G, Hobday AJ, Spaulding M, Gard M, Irons G (2019) Citizen reporting of wildlife interactions can improve impact-reduction programs and support wildlife carers. *Wildlife Research* **46**, 415–428. doi:10.1071/WR18127
3. Taylor-Brown A, Booth R, Gillett A, Mealy E, Ogbourne AM, *et al.* (2019) The impact of human activities on Australian wildlife. *PLoS One* **14**(1), e0206958. doi:10.1371/journal.pone.0206958
4. Camprasse ECM, Klapperstueck M, Cardilini APA (2023) Wildlife emergency response services data provide insights into human and non-human threats to wildlife and the response to those threats. *Diversity* **15**, 683. doi:10.3390/d15050683
5. Kwok ABC, Haering R, Travers AK, Stathis P (2021) Trends in wildlife rehabilitation rescues and animal fate across a six-year period in New South Wales, Australia. *PLoS One* **16**(9), e0257209. doi:10.1371/journal.pone.0257209
6. Commonwealth of Australia (2021) 'Australia State of the Environment Report 2021.' Department of Agriculture Water and the Environment, Canberra, Australia.
7. Soanes K, Lentini PE (2019) When cities are the last chance for saving species. *Frontiers in Ecology and the Environment* **17**(4), 225–231. doi:10.1002/fee.2032

8. Carter NH, Linnell JDC (2016) Co-adaptation is key to coexisting with large carnivores. *Trends in Ecology & Evolution* **31**, 575–578. doi:10.1016/j.tree.2016.05.006
9. Lute M (2019) *Coexistence: living harmoniously with wildlife in a human-dominated world*. International Fund for Animal Welfare, Washington, USA.
10. Soanes K, Sievers M, Chee YE, Williams NSG, Bhardwaj M, *et al.* (2019) Correcting common misconceptions to inspire conservation action in urban environments. *Conservation Biology* **33**(2), 300–306. doi:10.1111/cobi.13193
11. Van Helden BE, Close PG, Stewart BA, Speldewinde PC, Comer SJ (2021) Critically Endangered marsupial calls residential gardens home. *Animal Conservation* **24**(3), 445–456. doi:10.1111/acv.12649

Part 1: House

These lesser long-eared bats will not damage the electrical cables they are roosting next to.

Bats in the roof

SMALL, INSECT-EATING BATS MAY LIVE unnoticed in houses, sheds and other buildings for years. You may only realise you have microbats as housemates when you hear their chattering vocalisations, or perhaps notice their mouse-like scats. Another sign of resident bats is regularly finding a bat flying about the house.

LOCATION

Australia-wide.

SEASON

Temporary roosts are used all year round, with maternity roosts used in spring and summer, and winter roosts used to escape the cold.

SPECIES

The microbats that may be found roosting in buildings vary from state to state, and may include wattled bats, long-eared bats, free-tails and forest bats.

PHOTO: NIEL WARK

Behaviour

Bats in buildings

Unlike the generally much larger flying-foxes or fruit bats, which are known as megabats and eat pollen, nectar and fruit, microbats eat mostly insects. Microbats are often associated with living in caves, but 60% of all Australian bat species roost in tree hollows.[1] Tree-roosting bats are more likely to be found in and around our houses and sheds. The microbats that roost in caves are more likely to be found in concrete tunnels and under bridges. The microbats that may be found roosting in buildings vary from state to state, but broadly include:

- wattled bats (*Chalinolobus*), which have puppy-like jowls like spaniels.
- long-eared bats (*Nyctophilus*), which look like adorable little goblins. When they are sleeping or cold, long-eared bats may have their ears folded like a concertina fan; when they are active, warm and ready to fly their long ears are up and extended.
- most of the smaller free-tail species (*Mormopterus*), which have a small tail poking out from their wing tail membranes.
- the forest bat group (*Vespadelus*) and the small broad-nosed bat group (*Scotorepens*), which have small ears and sweet little faces with fine features.

The use of roost sites in microbats varies whether the bat is male or female, breeding or non-breeding, and also depends on the availability of food nearby and the overall availability of roost sites in the landscape.[2]

There are two kinds of roosts that are particularly important in a microbat's life: temporary roosts and maternity colonies.

Temporary roosts

Temporary roosts are smaller sites occupied by one or several bats. A microbat may have dozens of these temporary roosts scattered over their home range. Long-eared bats, for example, often shelter under loose bark, particularly the abundant bark on large old trees. In a study of two microbat species in a farming landscape near the Barmah Forest in Victoria, both species shifted roosts frequently: on average, individual lesser long-eared bats and Gould's wattled bats moved every 2–3 days.[2]

Temporary roosts have several advantages over permanent roosts such as caves, which can become known to predators such as pythons or birds of prey who wait at cave entrances for the bats to emerge at dusk. Changing roosts regularly avoids these lie-in-wait ambushes.

This group of broad-nosed bats (*Scotorepens* genus) was found roosting in a closed up garden umbrella. These tiny bats are vulnerable to predators, so leave the umbrella closed during the day. The best option is to let the bats move on in their own time, but if you want to use the umbrella, open it up at night while the bats are out hunting, and leave it open for a few days to discourage them from returning.

ROBIN ROWLAND

A temporary roost also minimises the accumulation of parasites built up in roosts[3] and allows the bat to select just the right microclimate for their needs.

Different roosts may be used depending on the weather conditions; for example, Gould's long-eared bats use exposed sites such as under a verandah ceiling during wet weather. In the southern states, microbats use their roosts to escape the cold and lack of insects in winter, where they go into torpor (a kind of light hibernation) until insects emerge in spring. Other temporary roosts around the house and garden include cracks in between wall joints, roller doors, furled pool umbrellas and even coats that are hung up in the shed.

Maternity colonies

The maternity colonies of tree-roosting bats are often in deep cracks in the trunk or hollows of eucalyptus trees. These old hollow-bearing trees can be removed for urban development, and those that remain in our parks and gardens are also often removed or trimmed due to safety concerns. Hollow-bearing trees can also be destroyed in planned burns, even if efforts are made to avoid this. These pressures have led to a lack of hollows across the landscape, and because these maternity colonies may be used by hundreds of bats year after year, in response some species have managed to establish roosts in people's roofs. A large colony of bats in a house or shed indicates that there is a good supply of insects in the area but a lack of natural tree hollows.[1]

The microbat breeding season varies a little from north to south, but generally the females give birth to one or two young in spring, and care for them until they are weaned a couple of months later, sometime in early summer. The males are not involved with rearing the young, and will be off in other roosts, sometimes alone or with other males, or even roosting with other bat species.[4]

Maternity colonies vary in size with some species such as Gould's wattled bat establishing large, almost permanent colonies in ideal conditions, such as a house or shed roof, while other maternity colonies are just as small and ephemeral as the temporary roosts. In the previously mentioned Barmah Forest study, a female lesser long-eared bat with young twins used eight different roost sites over an 11-day period – with six sites occupied for just a single day! This means many nights the mother bat moved one pup first, and then the other into the new roost![2]

Risks

To bats: Pet cats are one of the biggest predators of microbats in urbanised areas.[5,6] Bats such as lesser long-eared bats flutter around grass tussocks and shrubs, and even drop to the ground to pick up a moth or beetle, and are preyed upon by pet cats.

Microbats also suffer from collision with glass such as glass fences around pools and, in the tropical parts of Australia, collisions with fans.

To people/property: You cannot get any disease from bats simply living in or near your house. A small percentage of bats carry Australian bat lyssavirus, a rabies-like disease, which can only be transmitted via saliva from an infected bat. This can be avoided by not handling bats at all or by wearing gloves if you must touch a bat.[7]

Microbats will not cause damage to wood, or any internal fittings such as electrical cables or insulation, as they have no need to chew or modify their roost space.

Actions and solutions

If you have bats living in a section of your house, the first step is to get to know them. Do they live in their chosen roof cavity or crevice all year round, or just seasonally? If your bat colony is a year-round roost, or a maternity colony, leaving the bats where they are is always the best option. A large colony can create a lot of droppings and urine, so the best approach is to keep the bat colony quite separate to the living areas of your house. This will prevent the droppings or urine impacting on your living space, or the occasional lost bat inside. The sound of their chittering and chattering may also be less noisy if walls and ceilings are sealed properly.

You can talk to a carpenter to find out how you can best block any gaps between the interior and ceiling space or wall cavities, or do it yourself by using cloth or paper. Gaps can also be sealed with gap filler. Bear in mind that microbats can squeeze through the tiniest gaps - common areas include next to chimneys, exposed beams in cathedral ceilings and the gaps between ceiling lining boards.

Next, determine what entry and exit points they use. They may be using several, so you can engage friends or family to station themselves with different vantage points of the house and roof to watch the bats' dusk emergence simultaneously. This will also help you gauge the size of the colony, and their entry/exit points.

If you have no option and the bats must be excluded from the roof, professional advice from a wildlife rescue organisation, wildlife rescue consultant or bat ecologist is always recommended. If exclusion is the only option, then timing is very important to avoid harm to the bats. In the southern states, winter must be avoided as the bats may be in torpor. Spring is the peak time for young to be born, so exclusion activities must wait until the young have left the roost, which is mid-February in the southern states and late March in northern Australia. A fact sheet with instructions on how to create a plastic, one-way flap system that may be used to ensure all the bats leave and cannot re-enter can be downloaded from the Australasian Bat Society website.[8]

If your bats are using your house and garden buildings as temporary roosts, or you have a maternity roost, you can help by providing a variety of nest boxes affixed to trees around the house, or even on the house itself. This may encourage them to use the boxes instead of your buildings, but it may take a year or two for the bats to move in. Ideally place the nest boxes as high as you can, at ~4–6 m, as bats prefer them very high. This will require a ladder and a person willing to maintain the bats' new homes Nest box maintenance is not too onerous with those that are made well and from quality materials. It may involve reattaching the box if it comes down in wild weather, repainting periodically and removal of the box if unwanted species such as common mynas or bees move in.

What to do if you have a bat trapped in the house

If you have a bat flying around the living space of your house at night, you can use a combination of closed and open doors and windows to funnel it outside. For example, if the bat is flying around your lounge room, open all the windows of the lounge room and close the doors to any adjoining rooms. The expression 'blind as a bat' is a complete misnomer – most bats can see very well indeed! Your visitor will be using a combination of echolocation and vision to navigate through the house and should find its way to an open door or window eventually.

If a bat is trapped in your house during the day, it is preferable for them to stay inside until night falls, as birds such as currawongs and kookaburras will readily prey upon a microbat lost in the daytime. If the bat is hidden high in a curtain, for example, then you can safely leave them there. However, if you have a pet such as an indoor cat or curious dog, then for their own safety the bat should be caught using gloves and a pillowcase, and can then be kept in a small, ventilated cardboard box until nightfall.

If the bat seems unresponsive or unwell, call a wildlife rescue organisation for advice after capture. Unresponsiveness can indicate simply that the bat is in torpor, but it's best to seek advice to be sure.

To prevent bats becoming lost in your living space, try to find out where the bats are coming to the house. Bat entry points on houses can sometimes be located by the droppings stuck on the wall near the entry point and staining around the entry point from body oils. For example, you might have bats living in the wall. A small amount of staining near the top of a window frame indicates that they are gaining entry to the living space from a gap on the inside. Once this gap is fixed, the bats can go on living in the wall cavity without further accidents.

Identification tip: To determine whether small dark droppings are from mice or bats – do the crumble test! Bat droppings consist of tiny fragments of insect exoskeleton which break up easily when lightly crushed and rolled between your thumb and forefinger. Mouse droppings do not crumble, and also have a distinctive mouse smell.

Microbat friendly house tip: If bats are colliding with your glass pool fences, you can make the fences more visible by following the advice for birds on pp. 20-21. To avoid microbat collision with fans, turn off your outdoor fans each night when you go to bed.

References

1. Richards G, Hall L (2013) *A Natural History of Australian Bats: Working the Night Shift*. CSIRO Publishing, Melbourne.
2. Lumsden LF, Griffiths SR, Silins JE, Bennett AF (2020) Roosting behaviour and the tree-hollow requirements of bats: insights from the lesser long-eared bat (*Nyctophilus geoffroyi*) and Gould's wattled bat (*Chalinolobus gouldii*) in south-eastern Australia. *Australian Journal of Zoology* **68**, 296–306. doi:10.1071/ZO20072
3. Godinho LN, Cripps JK, Coulson G, Lumsden LF (2013) The effect of ectoparasites on the grooming behaviour of Gould's wattled bat (*Chalinolobus gouldii*): an experimental study. *Acta Chiropterologica* **15**, 463–472. doi:10.3161/150811013X679080
4. Bender R (2011) Bat roost boxes at Organ Pipes National Park, Victoria: seasonal and annual usage patterns. In *Biology and Conservation of Australasian Bats*. (Eds B Law, P Eby, D Lunney and L Lumsden) pp. 443–459. Royal Zoological Society of New South Wales, Mosman.
5. Legge S, Woinarski JCZ, Dickman CR, Murphy BP, Woolley LA, *et al.* (2020) We need to worry about Bella and Charlie: the impacts of pet cats on Australian wildlife. *Wildlife Research* **47**, 523–539. doi:10.1071/WR19174
6. Oedin M, Brescia F, Millon A, Murphy BP, Palmas P, *et al.* (2021) Cats (*Felis catus*) as a threat to bats worldwide: a review of the evidence. *Mammal Review* **51**(3), 323–337. doi:10.1111/mam.12240
7. Wildlife Health Australia (2019) 'WHA fact sheet: Australian bat lyssavirus'. March 2019, <https://wildlifehealthaustralia.com.au/Resource-Centre/Fact-Sheets>
8. Australasian Bat Society (2018) 'Bats in your belfry?' Fact sheet. Australasian Bat Society Inc., <https://www.ausbats.org.au/bat-fact-sheets.html>

A green tree frog in a suburban backyard, Sunshine Coast, Queensland.

There's a tree frog in my toilet

GREEN TREE FROGS ARE THE largest tree frogs in Australia, so it is certainly noticeable when one (or several) appear in your toilet. For some people, this can be a bit confronting in such an intimate context. You can encourage your guests to find other accommodation by building them a hotel.

LOCATION

Northern Territory, Queensland, New South Wales, Western Australia, South Australia.

SEASON

In dry regions such as inland Queensland and Northern Territory, frogs may enter your toilet in high summer or periods of drought. In tropical and subtropical areas, green tree frogs can appear in your house at any time of year.

SPECIES

The green tree frog, red tree frog, northern laughing tree frog, western laughing tree frog and Peron's tree frog.

PHOTO: ETHAN MANN

Behaviour

Green tree frogs are one of Australia's most well-known and iconic frog species, with their large size, expressive eyes and wide mouths that sometimes appear as if they are smiling. They are mostly a classic green colour, with some individuals more of an olive-yellow shade.

It is not only their distinctive appearance that makes them well known – they also take readily to our backyard habitats and even our homes. After a heavy rainfall green tree frogs can be found in all sorts of places – bathroom, toilet, drainpipes and even your letterbox. Some suggest that the frogs like the resonant qualities that these structures have for the frogs when they call – a low pitched 'werk werk werk'. In monsoonal areas such as Darwin, green tree frogs will enter your toilet during the dry season as they are seeking water.

Green tree frogs will take shelter under houses and verandah eaves during heavy downpours, and use walls, windows and balcony balustrading as handy vantage points for hunting moths and other prey. Tree frogs have a varied diet which is mainly insects and spiders, but also vertebrates such as lizards, other frogs and even microbats. When eating larger prey, frogs will use their forelegs to cram their mouths full!

In the first year of the Australian Museum's national FrogID project, green tree frogs were the most commonly recorded frog in Darwin, and the third most commonly recorded frog in Brisbane.[1] They are regarded as being reasonably tolerant of human-induced changes in the environment. Their use of human structures and gardens, their wide-ranging diet as well as their low-pitched calls are thought to contribute to their success in towns and suburbs. Low-pitched calls are important for any frog persisting in urban areas as their frequency is low enough to compete with suburban traffic noise.[2] But even though these frogs are regarded as tolerant, we cannot afford to be complacent. Green tree frogs used to be seen often in Sydney backyards, and researchers suspected that there was an ongoing decline in this previously common species. Citizen scientists contributing to the FrogID project confirmed this decline in 2017–18 with a mere 52 sightings of green tree frogs out of 7,120 recordings in the Greater Sydney area.[1] So, if you have a green tree frog in your toilet in Sydney – roll out the red carpet for your guest! And make sure you record any frog calls with the Australian Museum's FrogID app.

Other tree frogs that may be found in houses

The green tree frog (*Litoria caerulea*) is found in northern Western Australia, northern Northern Territory, north-eastern South Australia, and most of Queensland and New South Wales. Other tree frogs that may be found in our toilets include:

- *Litoria rubella*, red tree frog. Found throughout all of Queensland, the NT, the northern half of WA, most of NSW and northern SA.
- *Litoria rothii*, northern laughing tree frog. Found from the Mt Isa and Normanton regions, east to the Cape York region and then south to near Brisbane, Qld.
- *Litoria ridibunda*, western laughing tree frog. Found in northern WA and the Top End of NT, east to the Mt Isa and Normanton regions of Queensland.
- *Litoria peronii*, Peron's tree frog. Found in South East Qld, coastal and inland NSW, the ACT, south-east SA and most of VIC.[3]

Risks

To frogs: Cities and suburbs are harsh places for frogs. The development of urban areas often sees the removal of essential wetland habitat, with increased light, chemical and noise pollution plus direct threats to frogs such as traffic and pets. Unsurprisingly, an assessment of 196 frog species recorded on FrogID found that the species diversity of frogs reduces steadily as urbanisation intensifies.[4]

Your reasonably robust frog visitors can be helped by keeping pets inside at night when frogs are most active.

To people/property: None.

Actions and solutions

First of all, if there are frogs in your toilet it is perfectly fine to go on using it. If you don't mind, they don't mind! Most of the time the frogs cling to the underside of the toilet rim and get a vigorous shower of fresh water each time the human waste contents are flushed away.

While the frogs can handle ordinary toilet water, they are very sensitive to chemicals and artificial colours, so remove any freshener products immediately and switch to gentler cleaning products.

It is unlikely that the frogs will stay in your toilet permanently, so if you can, just leaving them alone until they return outside is the most frog-friendly option.

A frog's skin is a delicate and vital organ for the frogs as they drink and breathe through their skin! If you need to move a frog for any reason, it is vitally important that you wear wet gloves or at least have wet, clean hands as dry surfaces can tear their sensitive skin. Using gloves is preferred as frogs are susceptible to disease (see 'Frogs in trouble').

Frogs in trouble

Have you noticed dead or dying frogs at your place? Normally we don't see dead frogs, as they are often hidden away in dense habitat and decompose rapidly. In winter 2021, frog researchers, land managers, wildlife carers and the frog-loving Australian public in general were shocked to see reports from all over Australia of emaciated frogs outside in the daytime, either lighter or darker in colour than normal, with dry, peeling skin and dying in great numbers! This mass mortality event slowed in summer, then started up again in subsequent winters. The potentially deadly amphibian chytrid fungus (*Batrachochytrium dendrobatidis*), which causes the disease chytridiomycosis, is playing a part in these mass mortality events, but researchers suspect chytrid fungus is not the sole cause. You can help unravel the mystery by sending any reports of sick or dead frogs to the Australian Museum's citizen science project FrogID via calls@frogid.net.au. Include your location and, if possible, photos of the frog(s). Continued use of FrogID, recording frog calls as often as possible, also helps inform the impact this frog mortality event has had on Australia's frog populations. The research is ongoing and frog call submissions through FrogID can help.

Frog hotels

Frog ponds have long been a classic recommendation for the wildlife-friendly garden. These garden features require a bit of space, but even those with a postage stamp size courtyard or balcony can erect a frog hotel – specifically for tree frogs.

For a frog or a few frogs to take up residence in your toilet bowl, they need to navigate long pipes or perhaps take long journeys to access your bathroom. This may indicate the frogs' need for moist habitat.

Frog hotels are essentially bamboo or PVC pipes standing upright in a large pot, with gravel and sometimes water loving plants. The frogs take shelter in the shade and coolness of the pipes and enjoy the damp conditions created by the water wicking up the pipes from the gravel. Their sticky toe pads enable them to climb up the pipes and stick to the insides of their new home. See 'How to make a frog hotel' for a step-by-step guide how to build a frog hotel. If you don't have frogs in your toilet, frog hotels are still a great idea!

The frog hotel concept can also be used to encourage frogs out of your toilet. Instead of a large pot or tub, use a bucket with a handle. Assemble the pipes and gravel in

A green tree frog checked in to this newly created frog hotel soon after its creation.

ASHLEIGH MILLER

the same way and one-third fill the bucket with water. Place the bucket near the toilet, and then, as the frogs take up residence, you can pick up the bucket and move it gradually further away and then out of the house.

Other tips

You can cover your plumbing entry valves with fine mesh to stop the frogs accessing your toilet or bathroom from the outside.

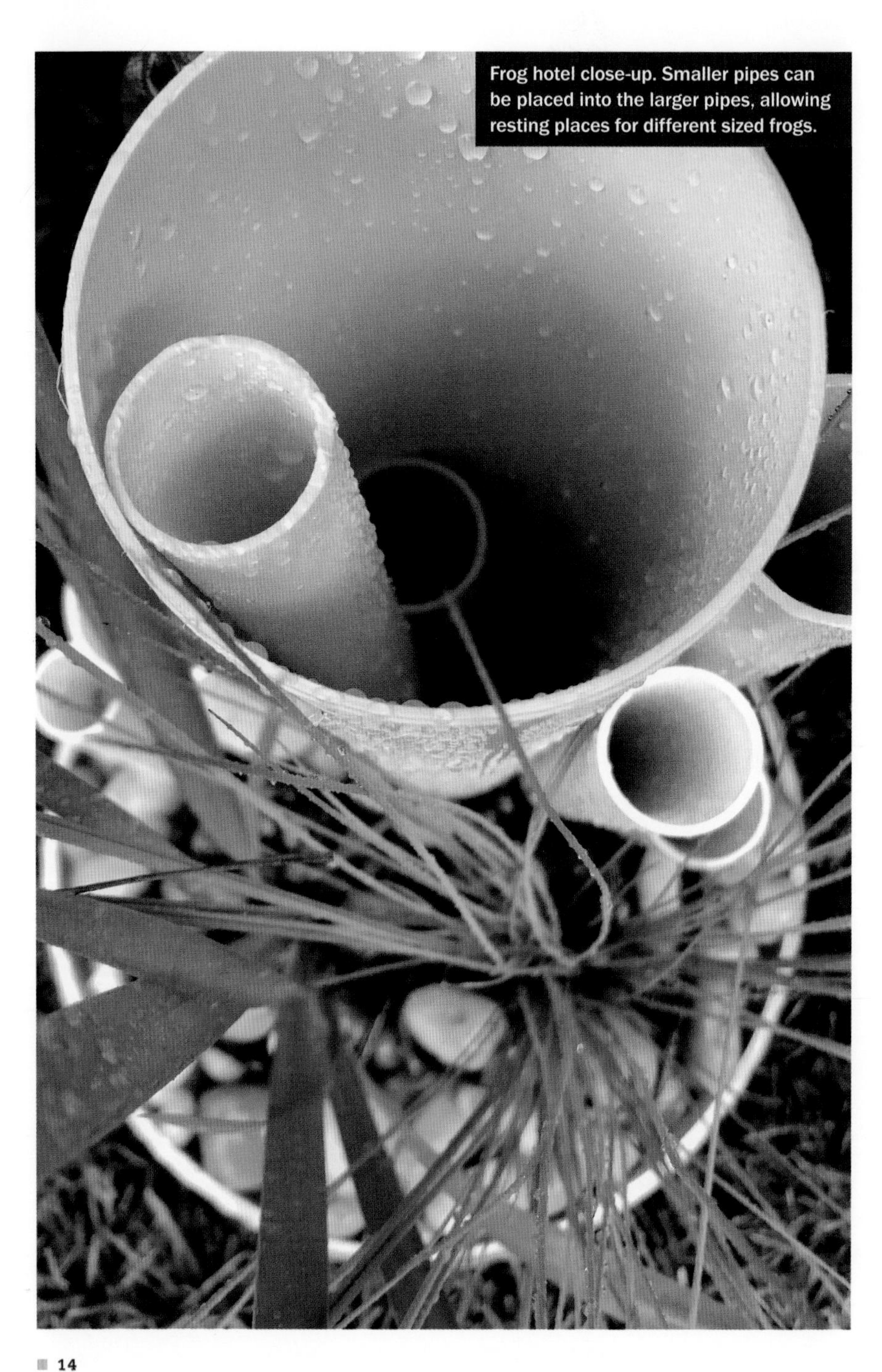

Frog hotel close-up. Smaller pipes can be placed into the larger pipes, allowing resting places for different sized frogs.

ASHLEIGH MILLER

How to make a frog hotel

YOU WILL NEED:

- **garden pot or vessel that does not have drainage holes**
- **2–6 bamboo or PVC pipes of various lengths and thickness**
- **sanding sheet, and saw or angle grinder**
- **gravel, pebbles and native potting mix**
- **water – preferably rain water. Or fill a bucket with tap water and let it sit for a few days.**
- **local water tolerant plants (if using).**

Sand the edges of the pipes to ensure they are smooth.

Place your container in its final location – cool and shady are best.

Add a layer of gravel in the bottom.

Arrange the pipes vertically in the pot, sometimes having a friend to hold them is handy!

Fill the space around the pipes with gravel and some potting mix (if using) until they are supported.

Arrange plants around the pipes.

Fill the pot and pipes with water, making sure water level stays under the gravel.

To make sure cane toads do not move into your green tree frog hotel make sure the pot is at least 60 cm high.

References

1. Rowley JJL, Callaghan CT, Cutajar T, Portway C, Potter K, *et al.* (2019) FrogID: citizen scientists provide validated biodiversity data on frogs of Australia. *Herpetological Conservation and Biology* **14**, 155–170.
2. Liu G, Rowley JJL, Kingsford RT, Callaghan CT (2021) Species' traits drive amphibian tolerance to anthropogenic habitat modification. *Global Change Biology* **27**(13), 3120–3132. doi:10.1111/gcb.15623
3. Frog ID (2020) *FrogID*. Australian Museum, Sydney, <http://www.frogid.net.au> (accessed 24 November 2023).
4. Callaghan CT, Liu G, Mitchell BA, Poore AGB, Rowley JJL (2021) Urbanization negatively impacts frog diversity at continental, regional, and local scales. *Basic and Applied Ecology* **54**, 64–74. doi:10.1016/j.baae.2021.04.003

The rapid flight of rainbow lorikeets and love for our backyard plants makes the birds susceptible to window collision

Birds hitting my windows

WE SEE A WINDOW FRAME and we know that a barrier is present. Window collision occurs because birds cannot see glass like we do, or they mistake reflected vegetation for the real thing. Happily, there are simple steps you can take to make reflective surfaces safer for birds.

LOCATION

Australia-wide.

SEASON

Collision rates peak during breeding season (spring) but do happen all year round.

SPECIES

Hundreds of bird species collide with windows, from the more abundant species such as the rainbow lorikeet to the very rare, such as the Carnaby's black-cockatoo.

PHOTO: ANTONINO NICOLACI

Behaviour

A flock of rainbow lorikeets chatters and shrieks, flying rapidly through the trees, shrubs and houses of suburban Sydney. Their excellent vision processes their surroundings rapidly, enabling them to make minute calculations and then execute precise turns with their muscles and feathers. But lorikeets and other birds see quite differently to humans. We have forward-facing eyes with binocular vision, whereas birds have lateral vision, with their eyes located further away from each other and on the sides of the head.

As they fly, the lorikeets rely on their lateral vision to monitor the behaviour of their flock members, look out for foraging opportunities, and for predators such as brown goshawks. Their lateral vision is more acute than their forward vision. On top of this, birds often turn their heads to look to the side or downwards as they fly, leaving them blind to the direction of travel.[1]

This forward-facing blind spot also explains why birds collide with chainlink fences and powerlines that seem so obvious to our human eyes. Bats also collide with windows (microbats) and glass pool fences (flying-foxes). The measures listed below will also help bats see glass.

A combined review of historical records (such as museum specimens) and more recent data from wildlife rescue organisations found 269 bird species have been recorded colliding with windows in Australia.[2] But the most recorded species involved in these incidents are the birds that live where we live – successful, urban-adapted birds such as the rainbow lorikeet, crimson rosella and Australian magpie. Threatened species such as Carnaby's black-cockatoo, forty-spotted pardalote, powerful owl and the critically endangered swift parrot also collide with windows. It has been estimated that each year up to 2% of the swift parrot breeding population – which is down to a mere 500 or so birds – is killed as a result of collisions with windows, fences (especially chain-link fences) and vehicles.[3,4]

Window collision also occurs when reflective glass creates mirror images of vegetation, which confuses birds into thinking they are flying towards shrubs or trees. Birds such as sacred kingfishers, laughing kookaburras and various pigeons fly quite differently to rainbow lorikeets and other parrots – with periods of inactivity and then bursts of rapid flight. These species are commonly recorded hitting windows in areas with rapidly growing urban encroachment as the suburbs spill into the surrounding bushland.[2]

Houses with a glass atrium or walkway between two living areas, and rooms with corner windows, also result in window collision as the transparent glass appears to provide a visual pathway.[5]

Sacred kingfishers are frequent victims of window collision.

NALINI SCARFE PHOTOGRAPHY

Large areas of transparent or reflective glass increase the risk of window collisions. Unfortunately, large expanses of north-facing glass are a key component of solar-passive house design.

Risks

To birds: The results of collision range from mild concussion to death. Window collision affects birds differently depending on the size of the bird, and the speed at which they were flying. Rates of window collision increase when birds are suffering from other factors such as disease and malnutrition.[2]

Most of the studies on window collision and mortality rates are from the United States and Canada where estimates suggest over a billon birds per annum are killed as a result of colliding with windows. The numbers are staggering, with an estimated 365 to 988 million deaths per year in North America[6] and 25 million birds in Canada.[7]

One surprising complication is that if a homeowner has a cat or a dog, then these death rates increase further, as birds are preyed upon by domestic pets under the window. One study in Argentina estimated that the rough annual bird mortality rate due to predation following bird-window collisions could reach ~6 million birds in Argentina (range = 1-11 million birds).[8]

'She flew away unharmed – or did she?'

My first experience with window strike was memorable. Early one morning there was a resounding crash and tinkle of glass in the loungeroom. I poked my head downstairs, and our dogs were quietly stunned, staring at the large female brown goshawk who was sitting on the couch surrounded by glass.

We covered her in a towel and moved her into a box to stay in the study for a few hours. Following the standard advice, we opened the box after this time, and we believed that 'the goshawk made a full recovery and flew off unharmed'. However now, over 15 years later, I wonder – did she fly off unharmed? I did not have the skills to make an accurate assessment of her injuries, and indeed those injuries may have been visible only by X-ray. If it happened again today, I would immediately call Wildlife Victoria and get a volunteer to transport the bird to either a vet or a shelter that specialises in raptors.

The scale of the problem is difficult to estimate in Australia due to the lack of reporting and standardised data collection. BirdLife Australia's Australian Bird Strike Project estimated that a minimum of 2.7 birds per day collided with windows across Australia in 2018, and over a period of 10 years (2009–19) the number of window collision reports dramatically increased each year.[2]

To people/property: Usually the window is fine – although older glass may shatter.

Actions and solutions

If the bird is sitting stunned beneath the window, use a tea towel or similar to pick up the bird and gently place it on a towel in the bottom of a cardboard box with air holes and a secure lid. Make sure the box is just large enough for the bird to extend their wings (not too big, not too small). Do not give it any food or drink. Close the box and put it in a quiet, dark and safe place. Call your local wildlife rescue organisation for advice.

Some wildlife organisations suggest letting the bird free after an hour or two of recovery. However, even if they become more active, do not let the bird go once you have contained them. Birds often have head and eye injuries, as well as internal injuries after hitting a window and may need help in the form of a vet check or period of rest at a wildlife shelter. Even minor injuries such as damaged feathers or bruising can prevent birds such as raptors from hunting effectively and they may die of starvation later. Being able to fly away is not an accurate indicator of the bird's health, however much we hope it to be.

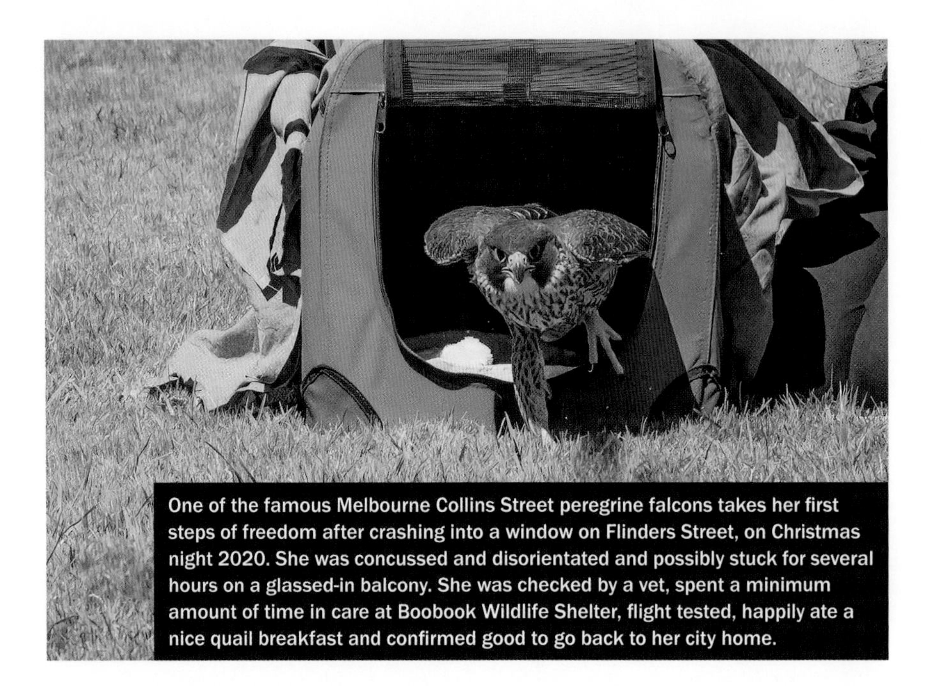

One of the famous Melbourne Collins Street peregrine falcons takes her first steps of freedom after crashing into a window on Flinders Street, on Christmas night 2020. She was concussed and disorientated and possibly stuck for several hours on a glassed-in balcony. She was checked by a vet, spent a minimum amount of time in care at Boobook Wildlife Shelter, flight tested, happily ate a nice quail breakfast and confirmed good to go back to her city home.

NALINI SCARFE PHOTOGRAPHY

Stopping window collision

Stopping window collision is a compromise between what we want as humans – to see our gardens, to let light in – and to modify our windows so that they are visible to birds in flight.

The most effective method to stop birds hitting your windows is to mark the outside of them with a series of vertical stripes at 10 cm intervals. The markings on glass can cover less than 7% of the window, whether the stripes are 2 mm thick or as wide as 13 mm thick they are equally effective.[9] Stripes can be applied to the window using tape or tempera paint.

Although vertical stripes are the most effective, dots, flowers, squiggles, medieval curlicues, smiley faces, even hawk silhouettes – all of these patterns work equally well as long as they are in a high contrast colour such as white, no more than 5–10 cm apart and cover the entire window. These patterns can be applied with a bar of soap, offering a cheap and temporary solution that is just as effective.

Another idea to modify your windows to prevent window strike is called a Zen curtain – a rod fastened to the top of the window with a series of strings at 10 cm intervals hanging down with a knot in the end. Zen curtains are visually attractive and therefore a great option for stylish homes or guest accommodation.

One common solution that is often presented is to cut out one or two large hawk silhouettes and stick them to the windows – the idea was that birds in flight would be deterred by the outline of a predator. But in 1989, one of the leading researchers in the field, Daniel Klem, demonstrated that the placement of single objects such as a bird of prey silhouette does not reduce strike rates to a significant level.[5]

Bird-friendly netting on the outside of the window will stop birds colliding with windows but is not visually appealing. External blinds that allow the human occupants to see out but cover the outside of the windows are another option, but these may be expensive and reduce visibility for the people.

References

1. Martin GR (2011) Understanding bird collisions with man-made objects: a sensory ecology approach. *The Ibis* **153**(2), 239–254. doi:10.1111/j.1474-919X.2011.01117.x
2. Aburrow K (2023) 'Bird Strike Project report 2020'. BirdLife Australia, Melbourne.
3. Hingston AB (2019) Documenting demise? Sixteen years observing the swift parrot (*Lathamus discolor*) in suburban Hobart, Tasmania. *Australian Field Ornithology* **36**, 97–108. doi:10.20938/afo36097108
4. Olah G, Waples RS, Stojanovic D (2024) Influence of molecular marker type on estimating effective population size and other genetic parameters in a critically endangered parrot. *Ecology and Evolution* **14**, e11102. doi.org/10.1002/ece3.11102
5. Klem D (1990) Collisions between birds and windows: mortality and prevention. *Journal of Field Ornithology* **61**(1), 120–128.
6. Loss SR *et al* (2014) Bird building collisions in the United States: estimates of annual mortality and species vulnerability. *Condor* **116**, 8e23.
7. Machtans CS, Wedeles CH, Bayne EM (2013) A first estimate for Canada of the number of birds killed by colliding with building windows. *Avian Conservation & Ecology* **8**(2), 19–33.
8. Rebolo-Ifrán N, Zamora-Nasca L, Lambertucci SA (2021) Cat and dog predation on birds: the importance of indirect predation after bird-window collisions. *Perspectives in Ecology and Conservation* **19**(3), 293–299. doi:10.1016/j.pecon.2021.05.003
9. Rössler M, Nemeth E, Bruckner A (2015) Glass pane markings to prevent bird-window collisions: less can be more. *Biologia* **70**(4), 535–541. doi:10.1515/biolog-2015-0057

A pale-yellow robin attacks his reflection furiously in the car mirror

Birds attacking their reflections

BIRDS ATTACK THEIR REFLECTIONS IN windows, car mirrors and even highly reflective surfaces such as glazed terracotta pots. The only solution is to remove the reflective nature of the surface. This can be a bit tricky, but it is worth it for the welfare of the animal.

LOCATION

Australia-wide.

SEASON

Window attacks peak during breeding season (spring) but do happen all year round.

SPECIES

Broadly includes magpies, kookaburras, robins, whistlers, magpie-larks, fairy-wrens, scrubwrens and even herons.

PHOTO: PATRICK TOMKINS

Behaviour

Often, this behaviour is mistaken for play, dancing or even admiring their reflections, but unfortunately the attacking bird is very stressed. The bird, usually the male, has his breeding hormones running high as he patrols his territory, gathers food and ensures his mate and nestlings are protected from intruders and predators.

After catching sight of his reflection in the window, he responds by singing his territorial call. Usually, a visiting bird would note the male's call and promptly fly away in order to avoid conflict. But when it is a reflection in the window, the other bird does not heed the warning or back down. The male adds a few more threat postures and the bird in the window responds in kind. The male has no choice but to take it to the next level – physical conflict! He attacks his reflection with his bill or feet, flying at the stubborn intruder. And he might do this all day, and for much of the breeding season.[1]

This behaviour was called 'shadow-boxing' by an un-named writer in the 24 May 1879 edition of the *Daily Times*, in New York. The article described an American robin that 'alights on the window ledge, taps vigorously on the pane, then flies up and down rapidly about three or four times'. Observers were mystified by the bird's behaviour, describing the bird's reflection as an evil spirit in bird form, an ominous bird, or that the bird was mistaking its reflection for a lost mate.

In 1894 American naturalist John Burroughs remarked on the behaviour and was unimpressed: 'It shows how shallow a bird's wit is when new problems or conditions confront it.' This is a little harsh, considering birds have been around for literally millions of years and reflective glass for less than 200 years!

Risks

To birds: This phantom attacking behaviour diverts precious energy from finding food for himself and his family and also distracts him from actual birds coming into his territory. The bird can become so incensed that he injures himself.[2]

To people/property: This behaviour is also stressful for the humans on the other side of the window – the attacks often start at dawn and are maddeningly repetitive. It is also distressing when one realises what a poor time the bird is having. Windows and mirrors may be damaged, and covered in droppings.

Actions and solutions

First, it is important to remember that not all birds do this all the time. Individual birds are in the right place at the right time to catch sight of their reflection and then they simply can't 'unsee' it.

Often the attacking is carried out from a handy vantage point, such as a hanging pot plant, a nearby water tank or the sprinklers on the roof that are part of the house fire protection safety system. Removing these perches is a good start but, once infuriated by the 'intruder', the attacking bird will most likely find another perch.

One method to help these stressed birds is cheap and relatively simple. Simply work out where the bird is seeing their reflection and hang light-coloured shade cloth over the outside of the windows, making sure you cover the entire window. After a month or so the breeding period finishes, and you can take the shade cloth down. Try removing one window covering first to see if the bird returns.

The author's kitchen window with a 'clay wash' a.k.a. dirt applied. All windows on one side of the house were treated and this stopped a grey shrike-thrush from attacking his reflection.

TANYA LOOS

There are other options to try:

- Soaping or dirtying the windows – with, for example, powder cleaners, soapy water or even dirty water. This method works particularly well for us as we have clay-rich soil. Simply mix a small amount of mud in a bucket of water and sponge over the windows. The fine sheer layer of mud can be transparent enough to let the light in, while stopping the reflection.
- Unfortunately, birds will start attacking multiple windows across the house – a total nightmare in houses with a lot of windows! Consider if there any windows you can sacrifice by applying a tinted or frosted covering. For example, you can use the frosted (non-reflective) window film that you can buy at a hardware store and that lets light in, but eliminates reflections.
- Another semi- permanent solution that modifies your windows but still allows you to look through them is a specially developed plastic film you apply to the outside of your windows that allows you to look outside, but completely eliminates reflection for the birds on the outside of the window. The film is available from US-based company CollidEscape.

Other tips

- If your glazed bird bath is very reflective at the base, wrap it in shade cloth.
- Remove all mirrors from your garden art.
- Birds attacking the car mirrors? Pop an old pillowcase or beanie on each mirror. Within a month or two, breeding season will finish and the pillowcases may be removed.

References

1. Roerig J (2013) Shadow boxing by birds – a literature study and new data from southern Africa. *Ornithological Observations* **4**, 39–68.
2. Dickey D (1916) The shadow-boxing of Pipilo. *The Condor* **18**(3), 93–99. doi:10.2307/1362511

This wombat appears healthy, with thick glossy fur and no mange visible.

Wombat living under the house

OCCASIONALLY A WOMBAT IS FOUND sheltering under the house, or even constructing a new burrow in between the foundations. For some people, especially wombat fans, this is a dream come true, but in some cases the wombat needs encouragement to move on.

LOCATION

South-east Australia, including Tasmania.

SEASON

Any time of year, but wombats may shelter under houses during extreme weather, such as flooding or bushfires.

SPECIES

The bare-nosed wombat and its three subspecies.[1] The bare-nosed wombat *Vombatus ursinus* used to be known as the common wombat, but 'bare-nosed' distinguishes them from the other two wombats in Australia – the northern and southern hairy-nosed wombats.

PHOTO: ISTOCK.COM/KEIICHIHIK

Behaviour

Wombats are one of the largest burrowing animals in the world, weighing up to 30 kg, with a formidable combination of strong musculature and broad paws with long digging claws. Wombats spend their days resting in safe, temperature-controlled burrows, emerging each night to graze on grasses, herbs, ferns and fungi.

This low-energy lifestyle means that wombats have comparatively small home ranges for their size -at just 10% of the size of a similarly sized wallaby or kangaroo![2]

Typically, a wombat will spend 1–4 days sleeping in the same burrow and then move to another burrow nearby. Burrows are most frequently occupied by just one wombat at a time. After one leaves, then another wombat uses the burrow. The exception to this rule involves the major burrows. Major burrows are used by generations of wombats and are very large, with multiple entrances and tunnels that can extend for 30 m. These wombat manors may have up to four individuals snoozing away simultaneously – albeit in quite different areas.[2]

Wombats are very solitary in nature, in fact one behavioural study of wombats in north-east Tasmania observed that 'on nine occasions individuals appeared to avoid coming close to each other purposely, either by changing their direction of movement or by running off in the opposite direction to an approaching animal'.[3] Wombats prefer to communicate via cube-shaped droppings left at prominent spots near the burrow entrances or on well-worn pathways, often accompanied by deep scratchings. These unique droppings are placed on higher points such as rocks or logs, and part of a wombat's evening commute to and from their chosen burrow involves inspection of these scat sites as the pheromones within contain vital information of recent burrow visitors, including their sex and breeding status.

Risks

To wombats: Bare-nosed wombats were previously widespread across southern Australia, but since colonisation their range has halved in size, and wombats on the mainland are now separated into two separate populations. The current population trend is unknown, but localised declines and extinctions occur when threats combine and result in population crashes – including habitat destruction, car strike and disease, especially sarcoptic mange (see 'Mange in wombats'). Wombats are also killed by humans – illegally and legally.[4] Around our houses, wombats are under threat from dog attack, but only large dogs. Cats who are allowed outside present an indirect threat to wombats, as their droppings may spread toxoplasmosis – a serious disease which affects wombats and other marsupials such as wallabies.[5]

Finn the wombat was orphaned when his mother died of toxoplasmosis. When he arrived at Bonorong Wildlife Sanctuary, he was severely underweight and needed weeks of intensive care to recover from the disease. Finn was released back into his natural habitat.

PETRA HARRIS – BONORONG WILDLIFE SANCTUARY

To people/property: Surprisingly, wombat occasionally attack humans. A friend of mine who is a wildlife rehabilitator heard a noise on her verandah one night, went to investigate and suddenly was set upon by a very large wombat, who bit both her legs severely. This is unusual, and in most cases, if you give wombats space – don't corner them, and back away if they start to charge – there will be no issues. Do not attempt to pick up an injured or unwell wombat. Wombat burrows can undermine house foundations.

Actions and solutions

Do a welfare check

Sometimes, a wombat will shelter under your house because it is displaced. If there is flooding or bushfires locally, a wombat's usual choice of burrows may be compromised. Roadworks or property construction can also displace wombats. The wombat could be sheltering under your house as temporary respite. During droughts, wombats may find areas under houses cooler than their usual burrows, and they may dig shallow sitting spots to rest in each night. These are usually temporary and not considered to be burrows.

Wombats may also shelter under houses when they are injured or diseased. Sarcoptic mange may cause blindness in affected wombats, and so wombats may shelter under houses if they are having trouble finding their way around. While you are conducting a visual welfare check from a safe distance, make sure you do not frighten the wombat or back them into a corner as they may charge, and can inflict some pretty serious injuries. Signs to look for are:

- obvious injuries, bleeding, laying on its side, not moving
- patches of fur missing from shoulders, back and head, eyes crusted over and rough scaly skin = mange.

Call a wildlife rescue organisation for advice. If your visitor appears healthy, leave them alone and let the wombat use your space until environmental conditions improve and burrows are available again.

Note that trapping and translocation of the wombat are not viable options. In South Australia, researchers have shown that in the closely related southern hairy-nosed wombats this method does not solve the problem. After spending considerable time and energy trapping individual wombats, the burrows at the study site were recolonised within 1-2 weeks.[6]

The permanent solution: wombat mesh

Let's say you are away for a few weeks, and you come home to a very active burrow under the house with lots of fresh diggings, scratchings and scats, and a determined wombat coming and going from their new favourite burrow.

Any large rocks, logs or other heavy obstructions will simply be moved aside by the determined wombats. The only way to evict a wombat from under the house is by the use of wire. The wombats do not like digging through wire so weldmesh or even lighter rabbit wire may be used. Lay the mesh around the entire house. You will need to place weldmesh or rabbit wire down from your house base either into the ground by at least 600–900 mm or down and then out as a 'skirt' by the same amount. If a wombat burrow is in use, you will need to create a one-way flap that allows the wombat to leave the burrow, but prevents their return the following night. This involves cutting a hole out of the mesh that is perpendicular to the house, then attaching the flap – a bigger piece of wire that is hinged over the opening allowing the wombat to leave.

Once your wombat has left, then make sure you have a mesh barrier on any exposed dirt between the house and the ground. This prevents your wombat, or another wombat, moving back in at another time.

Electric fencing placed over the area the wombats are entering under the house is another option,[7] although you need to make sure the wombat has definitely vacated the burrow before turning it on!

Mange in wombats

Sarcoptic mange is caused by the tiny mite *Sarcoptes scabiei*. The mite causes intense itching, and if left untreated leads to severe skin infections, blindness, starvation and eventually death. In humans the disease is known as scabies. It is believed that colonists brought the mite with them and their dogs – and in the last 200 years the mite has spread to introduced red foxes and native wildlife as well. The mite also affects koalas, wallabies and possums, but it is of most concern in the wombat population, as their use of burrows and slow-paced lifestyle are ideal for mite proliferation. A recent citizen science study using WOMSat sightings to report both healthy and manged wombats discovered mange wherever wombats occur.[8] Wombats are particularly susceptible to mange when they are stressed, possibly from high wombat densities, periods of drought or after flood. In these cases mange prevalence can be in half the population, with near 100% mortality.[5] Groups like the Wombat Protection Society and Mange Management are working closely with the wildlife rehabilitation community, veterinary scientists and government agencies to treat affected wild wombats.[9]

Other tips

Some sources suggest placing blood and bone and other smelly substances at the burrow to encourage it to move on. This is untested and, if a wombat really likes its new burrow, unlikely to work. Setting up a light and a radio at the burrow when you go to bed is another deterrent, but again not a permanent solution.

References

1. Martin A, Carver S, Proft K, Fraser TA, Polkinghorne A, *et al.* (2019) Isolation, marine transgression and translocation of the bare-nosed wombat (*Vombatus ursinus*). *Evolutionary Applications* **12**(6), 1114–1123. doi:10.1111/eva.12785
2. Evans MC (2008) Home range, burrow-use and activity patterns in common wombats (*Vombatus ursinus*). *Wildlife Research* **35**(5), 455–462. doi:10.1071/WR07067
3. Taylor RJ (1993) Observations on the behaviour and ecology of the common wombat (*Vombatus ursinus*) in northeast Tasmania. *Australian Mammalogy* **16**(1), 1–7. doi:10.1071/AM93001
4. Thorley RK, Old JM (2020) Distribution, abundance and threats to bare-nosed wombats (*Vombatus ursinus*). *Australian Mammalogy* **42**(3), 249–256. doi:10.1071/AM19035
5. Wildlife Health Australia (2021) 'Sarcoptic mange in Australian wildlife'. Canberra, <https://wildlifehealthaustralia.com.au/Portals/0/ResourceCentre/FactSheets/Mammals/Sarcoptic_Mange_in_Australian_Wildlife.pdf>.

6. O'Brien C, Sparrow E, Dibben R, Ostendorf B, Taggart D (2021) Translocation is not a viable conflict-resolution tool for a large fossorial mammal, *Lasiorhinus latifrons*. *Wildlife Research* **48**(1), 7–17. doi:10.1071/WR19195
7. Martin AM, Burridge CP, Ingram J, Fraser TA, Carver S (2018) Invasive pathogen drives host population collapse: effects of a travelling wave of sarcoptic mange on bare-nosed wombats. *Journal of Applied Ecology* **55**(1), 331–341. doi:10.1111/1365-2664.12968
8. Mayadunnage S, Stannard HJ, West P, Old JM (2023) Spatial and temporal patterns of sarcoptic mange in wombats using the citizen science tool, WomSAT. *Integrative Zoology*. Online early. doi:10.1111/1749-4877.12776
9. Mounsey K, Harvey RJ, Wilkinson V, Takano K, Old J, *et al.* (2022) Drug dose and animal welfare: important considerations in the treatment of wildlife. *Parasitology Research* **121**(3), 1065–1071. doi:10.1007/s00436-022-07460-4

A mother and joey common brushtail possum in a garden shed roof.

Possums living in the roof

A POSSUM LIVING IN YOUR roof can be a source of joy, but perhaps their den is just above your bedroom and the loud stomping across the roof is wearing thin. The good news is that, with some persistence, you can safely exclude possums from your roof without causing them undue harm.

LOCATION

Australia-wide.

SEASON

All year round.

SPECIES

Most likely the common brushtail possum, but can involve other species.

PHOTO: ISTOCK.COM/PLEIO

Typical backyard possums

The common brushtail possum (known as the northern brushtail in the Northern Territory) is most likely to take up residence in a roof cavity, but there are exceptions. In forested, higher altitude regions in Tasmania, New South Wales and Victoria, your roof possum may be a mountain brushtail possum, also known as the bobuck, and in tall forests and rainforests in New South Wales they could be the closely related short-eared brushtail possum.

In Albany, south-west Western Australia, roof possums are more likely to be western ringtail possums. This is surprising because ringtail possums can make their own shelter: a spherical nest made of finely shredded bark and leaves known as a drey. It is not clear why roofs are used by ringtail possums when they could make their own nest – perhaps the roof space offers better protection from predators, or the garden vegetation on site is not dense and tangled enough to support a drey.

The eastern counterpart of the western ringtail is the common ringtail. In the eastern states, ringtails are less likely to take up residence in roofs, but are common garden residents.

Ringtails are often thought of as 'baby possums', as the adults are small and very appealing. The key identifying features of ringtails (both western and common) are their small size and their tails, which have a distinctive white tip and are covered in short fur that is the same length or shorter than their body fur.

Brushtail possums, including common brushtail, northern brushtail, short-eared brushtail and mountain brushtail, are all quite large, about the size of a cat. Their black tails vary between bushy and more sparsely furred and might appear naked at the tip.

Behaviour

Brushtails can be overabundant in urban areas, living in our gardens and in our parklands in such numbers that you may often see large 'possum collars' wrapped around tree trunks to prevent over-browsing of tree foliage. One might think that the species is doing rather well, but out in their natural woodland and forest habitat it's another story – brushtail possums have become locally extinct over more than half their previous range.[1] Brushtail possums are now considered common only in some cities, Tasmania and islands such as Kangaroo Island and Magnetic Island.[2]

In bushland areas, brushtail possums spend the day in large eucalyptus hollows called dens. Brushtail and ringtail possums will use their den or drey during the day, and then sometime after sunset they will emerge to forage for food. Possums tend to travel regular routes each evening and again when they return early in the morning.

Brushtails may have several dens in different parts of their home range, but particular dens are used depending on the weather, if it is breeding season and the availability of food resources. The continuing destruction and loss of hollow-bearing trees in forest habitats, cities and suburbs is creating competition with other species, such as feral honeybees and rainbow lorikeets, and is further reducing natural hollow availability.

Brushtail possum home ranges differ between males and females. In one Tasmanian study, home ranges were up to 8 ha for males but just 1–2 ha for females.[3] These small home ranges (especially in urban areas) mean your garden or roof may be the site for a territorial dispute. This can sound quite alarming, with guttural growling calls and some shrieking and hissing. Ringtail possums are quieter and their soft, high-pitched twittering calls are not heard as often as brushtail possums.

Risks

To possums: Living with humans is quite risky for possums. The main reasons for possums coming into care are cat and dog attack, car strike, powerline electrocution and poisoning from rodenticide. They are also exposed to toxoplasmosis from pet cats. Brushtail possums in warmer parts of Australia such as New South Wales, Queensland and Northern Territory may succumb to stress dermatitis, which is an ulcerative skin condition that may result in death of the animal. The cause of the dermatitis is unknown, but it occurs when a scratch becomes infected, and may be more common in overcrowded or stressed possum populations.[4] Possums may also become trapped if they choose an unsuitable den site such as a chimney or above wood fires.

While some people love 'their' possums and enjoy feeding them, this can cause disease and altered behaviour (as the young do not disperse), and may lead to increased population density. An overabundant possum population can then lead to frustrated neighbours breaking state wildlife laws by renting traps to relocate possums illegally or even capture and kill them.[5]

To people/property: Noise disturbances from the possum and urine stains on walls.

Actions and solutions

Make sure it is a possum!

Noises in the roof are often assumed to be a possum, but are very often a quite different animal: the introduced black rat (*Rattus rattus*). These rats nest in roofs and can be quite noisy, making gnawing sounds and high-pitched squeaking, whereas possums growl, hiss and shriek.

The easiest way to identify your roof inhabitant is to find their droppings or scats. Possum scats are distinctive, being the largest (usually more than 2 cm long) compared to rat (~1.5 cm long) and mouse (less than 1 cm long) droppings (for more on rats in the roof, see pp. 49–56). If you can access your roof, you can also sprinkle flour on the roof cavity floor and check the footprints – possum footprints are the same size as those of a cat.

If your roof inhabitant is a possum and you live in Western Australia – congratulations! Your property is providing habitat for the western ringtail, an endangered species. In fact, governments in these states would love to hear about your possum, and they have booklets and other information for landholders who are hosting this species.

If possible, accept your possum

Whether you have a brushtail or a ringtail in the roof, the next step is to understand that the friendliest solution avoids possum removal or relocation from your property. Your property has provided the resources they need to survive, such as shelter in your roof, water and food (maybe even your vegetables or roses!). If you have all the elements of a possum-friendly garden and you relocate your possum, another one will simply move in.

A study of the fate of possums removed from houses in Melbourne revealed starkly that brushtail possums relocated to bushland nearby do not fare well at all – in fact over 70% of relocated possums died within a week.[6] The relocated possums are suddenly in unfamiliar territory, with no way of knowing locations of den sites, food or water. In a bushland area with suitable food and shelter, there may already be a resident population of possums. The study showed that relocated possums spent 68% of their time at ground level, including den sites – a dangerous and unfavourable situation. This study, prompted by Ian Temby of the Victorian State Government, resulted in changes in legislation, recognising that bushland relocation is an unacceptable solution for possums.

If exclusion from the roof must occur

If you must exclude your possum from the roof, you will need to work out where they are getting in and out. The entry and exit hole may be as small as the size of your fist, even if a larger possum species is in residence. A giveaway is a discoloured area around the gap, as the oils from the animal's fur eventually create a stain. Tufts of fur may also get caught at these points.

If you are having trouble locating the entry site, you can engage friends or family members to carry out a possum watch. With people stationed quietly at various vantage points around the garden, your possum will be spotted exiting their roof den soon after dark. Bear in mind that sometimes possums can travel through the wall from the roof and exit your house at ground level!

If you notice your possum has a baby or two carried on her back, try to put up with any possum problems for a few more months before attempting exclusion. Losing a safe and secure den at this point could be very stressful for the mother possum and her joey.

Once you know how the possum or possums are getting into your roof, you have a few options depending on how handy you are. You can build a special, one-way flap that allows the possum to get out but not back in again – this method is great as you can be sure that you avoid accidentally trapping an animal in your roof cavity. Alternatively, if you have seen your possum leave and feel confident to do so, you can block the holes with cheap, stretchy chicken wire, but remember this must be done at night.

Quassia chips and bright lights in the roof are said to discourage possums, but urban-adapted possums are habituated to all manner of human-created sights, smells and sounds, and these strategies are unlikely to have an effect on determined, 'street-wise' possums. Blocking access into the roof space is the only fail-safe method as it removes the possibility of your roof space becoming a den site.

If you can't locate the entry and exit points, or are unsure of how best to modify your roof to remove any gaps, you can employ a possum catcher (preferably an animal rescue leaning consultant rather than a 'pest' controller). Many of these possum controllers now use one-way flaps because it's easier than trapping and more humane.

Other tips

Avoid feeding possums and plant possum-friendly plants instead (see pp. 123–129). Possums will also take advantage of other food around the house, so take any pet food indoors at night and cover your compost heap or bin.

If you want to discourage possums from running over your roof at night, trim any branches that hang within 1.5 m of the gutter. Make sure any possums living in the roof have been excluded before the branch trim!

Invest in a possum box or two around your property. Your local wildlife organisation can advise you where to purchase nest boxes, or how to locate some possum box designs to build one yourself.

References

1. Goldingay RL, Jackson SM (Eds) (2004) *The Biology of Australian Possums and Gliders*. Surrey Beatty & Sons, Chipping Norton.
2. Isaac JL (2005) Life history and demographics of an island possum. *Australian Journal of Zoology* **53**(3), 195–203. doi:10.1071/ZO05018

3. Statham M, Statham HL (1997) Movements and habits of brushtail possums (*Trichosurus vulpecula* Kerr) in an urban area. *Wildlife Research* **24**, 715–726. doi:10.1071/WR96092
4. Eymann J, Herbert CA, Cooper DW (2006) Management issues of urban common brushtail possums *Trichosurus vulpecula*: a loved or hated neighbour. *Australian Mammalogy* **28**, 153–171. doi:10.1071/AM06025
5. Wilks S, Russell T, Eymann J (2013) Valued guest or vilified pest? How attitudes towards urban brushtail possums *Trichosurus vulpecula* fit into general perceptions of animals. *Australian Zoologist* **34**, 33–44.
6. Pietsch RS (1994) The fate of urban common brushtail possums translocated to sclerophyll forest. In *Reintroduction Biology of Australian and New Zealand Fauna*. (Ed. M Serena) pp. 239–246. Surrey Beatty and Sons, Chipping Norton.

A sulphur-crested cockatoo gnaws at a wooden rail with their formidable bill.

Cockatoos destroying my house

LOVEABLE LARRIKINS OF THE SKY, sulphur-crested cockatoos delight us with their spirited screeching and apparent zest for life. But when they direct their boundless enthusiasm towards the destruction of windowsills, balconies and children's play equipment, our opinion of them can sour.

LOCATION

East coast of Australia, and northern Australia. In Western Australia sulphur-crested cockatoos are an introduced population and do not occur in large numbers, and house or building damage is more likely to be from little corellas.

SEASON

All year round, with increased activity in spring.

SPECIES

Mainly the superbly urban-adapted sulphur-crested cockatoo, but corellas such as little corellas may also damage houses in rural areas.

PHOTO: ISTOCK.COM/KEN GRIFFITHS

Behaviour

You may experience frequent cockatoo visits to your garden for decades without any problem, and then one day return home to find your windowsills are damaged or the play equipment is shredded. The destruction can be breathtaking in its severity – and very expensive to repair! The situation can be doubly frustrating for homeowners as most home insurance companies have a clause excluding animal damage from 'pecking or biting'.

Sulphur-crested cockatoos live in a wide range of habitats, from rainforest to eucalypt forest and even the dry woodlands of the mallee. In these areas, cockatoos spend part of each day roosting in shady trees, relaxing with their flock mates in an activity technically termed loafing. While loafing about, the cockatoos often gnaw on the bark and wood of branches, helping to maintain the condition of their bills and ensure they do not overgrow. It is also presumably satisfying for the bird in the same way a dog enjoys chewing a fresh bone, as it exercises the jaw and passes the time.

Urban cockatoos still need to whittle away their excess bill growth, and the structures in cities and suburbs provide plenty of opportunities. Western red cedar on windows, doors, balustrades and outdoor play equipment are particularly vulnerable to cockatoo damage. As well as the more traditional wood substrate, urban cockatoos attack painted polystyrene moulding used in shopping centre buildings, light fittings and telecommunications cabling, and may spend time on the roof removing loose screws.[1]

In 2008, a flock of cockatoos took a liking to methodically smashing each of the thimble-sized light globes in the Arts Centre Melbourne's spire, causing over $70,000 worth of damage.[2] In 2010, strata managers of Sydney Campus Apartments claimed that close to $100,000 worth of damage had been carried out by cockatoos, and subsequently filed for a permit to cull 20 cockatoos. Only two were killed before the cull was halted thanks to public outcry.[3] More recently, cockatoos have been filmed carefully removing the metal anti-bird spikes along the ledge of a building in Katoomba, NSW.

Anti-bird spikes have also been removed by the little corellas in Geraldton, WA, where they have spent much time on the roof of Saint Francis Xavier Cathedral, chewing at metal fixtures, removing the lids off electrical junction boxes, and eating through wires.[4]

Sulphur-crested cockatoos and corellas are bold and curious, and the sheer range of items likely to be attacked is clearly beyond the need to chew wood to sharpen and shape their bills – there is likely to be an element of play and the enjoyment of problem solving involved. It's clear that anyone hoping to discourage cockatoos from chewing their house and garden equipment needs to be as wily and innovative as the cockatoos themselves!

Bin-opening cockatoos

As a testament to their intelligence and problem-solving ability, urban cockatoos in the Sydney region and at Lorne in Victoria have taught themselves how to open rubbish bins. An individual bird, most often a male, pries the lid open with his bill, then edges around the bin until he generates enough force to flick the lid open. The rest of the flock then descends and forages in the rubbish. From a scientific perspective, bin-opening by cockatoos is a wonderful opportunity to study intelligence and even the transmission of culture in a wild animal. But for affected local residents, although not as expensive as house destruction, this behaviour is extremely annoying.

Aided by the citizen science program Big City Birds and annual online surveys, researchers led by Barbara Klump from the Max Planck Institute and Lucy Aplin from the Australian National University have published several studies on this behaviour.[5] Cockatoos in different suburbs have different styles of opening the bins, showing they even have regional 'subcultures'. The team have also managed to document an ongoing battle of will between cockatoos and humans – as humans employ more methods to stop the cockies, the cockies use their wits and strength to overcome them.

While a brick or stone on the bin lid stopped the birds for a while, soon enough the cockatoos learned to push the object on the ground and open the bin as usual. Residents tried plastic owls and rubber snakes, which were ignored. At the time of writing, weighting the bin lid from the inside seems to do the trick, but residents joke that a padlock may be the logical next step!

A sulphur-crested cockatoo opens a bin lid while another individual watches with interest.

MAX PLANCK INSTITUTE OF ANIMAL BEHAVIOR/BARBARA KLUMP

Risks

To parrots: The greatest risk to cockatoos and corellas who damage buildings is calls for state-sanctioned culling by frustrated residents. However, this solution is ineffective as other parrots may carry out the same behaviour.

To people/property: Cockatoo rescuers and companion parrot owners know that a cockatoo's bite can be painful, but there is no danger to humans from cockatoos or corellas visiting your house and garden. Estimates of property damage vary. At a public meeting in Upwey, Vic, in 2008, a group of 61 residents reported a total of $241,000 worth of damage to their houses.[6]

Actions and solutions

Solutions can be grouped loosely into three areas: create an effective barrier, make it hard for the parrots to perch and chew, and convince the cockatoos that your home isn't a safe, relaxing place.

Create an effective barrier: Protect the damaged area by covering it with chicken wire or external shade cloth blinds that can be rolled down when you are out. Some sources suggest cockatoos do not like the colour white.

Make it hard to perch: Create a trip wire on balconies using wire or fishing line strung from nail to nail. A rolling perch, made from a piece of wooden dowel with a slightly larger piece of plastic pipe that rolls easily when a bird tries to land on it, can also discourage perching. These measures won't harm the parrot - they will use their wings for balance and fly off easily to find somewhere easier to perch.

Try deterrents: Spraying the birds with a water pistol or hose whenever the cockatoos land can be an effective and harmless solution, but this only works if you are home much of the time. If you have the funds, you can purchase a motion-activated, solar-powered spray repeller.

Playing the recorded calls of birds of prey is unlikely to work as a deterrent to your cockatoo visitors. Birds of prey such as goshawks hunt for birds using a stealthy and silent approach that takes their prey by surprise.

Prevention

Over 100 reports of cockatoo damage to houses nearly always involve someone feeding birds - either at the property, or in a backyard nearby - which can discourage them from moving on.[7] In short, do not feed your local cockies![8] If you are not feeding birds but are still suffering from cockatoo damage, you may need to make some queries and ask your neighbours to stop feeding wild parrots.

References

1. Temby I (2018) 'Guidelines for reducing cockatoo damage, wildlife management methods'. The State of Victoria Department of Environment, Land, Water and Planning, Melbourne.
2. Beck E (2008) Cockatoos not so rapt as Zorro flies in. *The Age*, Melbourne, 15 February.
3. Hanney R (2010) Cockatoo cull? *Tasmanian Times*, 25 October.
4. Temby I (unpublished) 'Managing impacts of the little corella on the Fleurieu Peninsula'.
5. Klump BC, Martin JM, Wild S, Hörsch JK, Major RE, *et al.* (2021) Innovation and geographic spread of a complex foraging culture in an urban parrot. *Science* **373**(6553), 456–460. doi:10.1126/science.abe7808
6. Martin T (2008) Beating the birds. *Star*, Ferntree Gully/Belgrave, 18 November.
7. Perrin DJ, Pullen BT, Cox GH, de Fegely RS, Evans DM, *et al.* (1995) 'Problems in Victoria caused by long-billed corellas, sulphur-crested cockatoos and galahs'. Environment & Natural Resources Committee, State Parliament of Victoria, Melbourne.
8. Jones D (2022) *Curlews on Vulture Street: Cities, Birds, People and Me*. NewSouth Publishing, Sydney.

A white-tailed spider pauses with their spider prey, in this case a wolf spider.

PHOTO: ALAN HENDERSON – MINIBEAST WILDLIFE

Spiders in the house

SOME SPIDERS FIND OUR HOMES ideal habitat with corners for web building, walls for hunting – and plenty of insects and other spiders to feed upon. Sharing your house with our eight-legged housemates is perfectly safe with a few simple precautions.

LOCATION

Houses, verandahs and sheds Australia-wide.

SEASON

Warmer months in southern Australia, all year round in warmer climates.

SPECIES

Daddy long-legs, black house, redback, white-tailed, huntsman and Sydney funnel-web spiders.

Behaviour

The behaviour of these spiders in our homes varies depending on their hunting strategy – whether web-based or ambush.

The most common species in houses is the daddy long-legs, which is in fact several species, with both introduced and native members in this family (Pholcidae). Other web builders found in our homes include the black house spider and the redback spider. Another infamous spider is the white-tailed spider, which does not build a web. Huntsman (Sparassidae) are famed for their huge size and are found in houses throughout Australia. These spiders are known as modern spiders, the araneomorphs. The Sydney funnel-web is a mygalomorph, a group of large primitive spiders which live in burrows. As the name suggests, the Sydney funnel-web only occurs in a 160-km radius around Sydney – but there are other primitive spiders that look very similar, such as mouse spiders and trapdoor spiders.[1]

The daddy long-legs spider catches their prey via a tangled, unstructured web, usually erected in the corner of a room, or behind a picture frame. They also create trip lines of silk that are nearly invisible, radiating out from the web and allowing the spiders to sense their prey over a much wider area.

The black house spider builds their web in the corner of a window, both inside and out, and around light fittings. The web is untidy and may have a cocoon-like tunnel where the spider shelters during the day. Away from houses, these spiders build their messy webs in the gnarled bark of huge eucalypts or in rocky under hangs.

The redback spider also has an untidy web, but prefers dark, rarely disturbed corners – most famously the outdoor 'dunny', but also under verandahs. We have had a redback spider in our letterbox, and also in the corner of our fire bunker. Redback spiders are not as common these days in warmer parts of Australia since the introduced Asian house gecko seems to like preying upon them![1]

The huntsman spiders are instantly recognisable with their very long legs and large size. Huntsman spiders are ambush predators – they do not build a web, but rather wait in a favoured spot to snatch insect prey. When frightened, these spiders can run at an incredible pace. The fastest huntsmen recorded are both from tropical Queensland – northern banded hunstman *Holconia hirsuta* and golden huntsman *Beregama aurea* – and can run at 42 and 31 body lengths per second, respectively.[2]

White-tailed spiders do not build a web – instead they move through their habitat, whether that be bushland or the downstairs rumpus room, preying upon insects and other spiders such as black house spiders.

Sydney funnel-webs and other mygalomorphs are burrow living, and rarely come inside the house. Males are the most commonly encountered as they wander in spring searching for females. These spiders hunt from their burrows, using trip lines to alert them to the presence of passing prey.

Risks

To spiders: The risks to spiders in the home very much depend upon the attitude of the humans present. Many people tolerate some level of spider population in the house while others are frightened of spiders and use pesticides or vacuum the spiders away, even while they are still alive!

To people/property: Spider bites and hospitalisations do occur. But compared to bees and wasps or even dog bites, the risk is minimal. Between 2017 and 2018 there were over 3,500 Australians in hospital due to contact with a venomous animal or plant, and more than a quarter of them were due to bee and wasp stings. Spider bites accounted for just one-fifth (19%) of hospital visits. For comparison, in the same year 9,542 people were hospitalised due to dog attack.[3]

Myths abound about the dangers of spiders, probably related to the prevalence of arachnophobia; fear of spiders may be the most widespread fear of animals.[4]

The daddy long-legs spider is said to have the most potent venom of any of the spiders, and the only reason they aren't featured in Table 1 is because their jaws are too small to penetrate human skin. This is completely untrue, and probably arose because people observed the fearsome redback spiders being caught and eaten by daddy long-legs – so the logic went, these must be powerful! But the truth is more mundane in nature – the daddy long-legs' legs are so long they can wrap up the redback without coming into reach of their fangs.[1]

White-tailed spider bites are said to cause a 'flesh-eating' wound much like gangrene. However, a survey of 130 bites by this species found no evidence of necrosis.[5] Instead, bites had minor effects, or at worst painful red lesions.

Sydney funnel-web venom is very potent indeed, but happily there have been no deaths since the development of an antivenom in 1981.[1]

Table 1. Spiders and hospitalisations in Australia 2017–18.[3]

Spider	Number of hospitalisations in 2017–18
Redback spiders	283 cases
White-tailed spiders	38 cases
Funnel-web spiders	25 cases
Unknown spider	300 cases

Actions and solutions

Your next steps depend on your tolerance level for spiders and their location in the house. Even the most ardent spider lover may find it difficult to turn out the lights and go to sleep when there is a large huntsman spider just above the bedhead.

- Daddy long-legs are harmless – they can be left where they are, and the webs cleaned away with a stick occasionally to prevent excessive build-up.
- Black house spiders stay in their webs, so if you don't mind the gothic décor, you can leave them *in situ*.
- Redback spiders are best removed and taken outside – use a stick to wrap both spider and web up, and wear gloves just in case!
- Huntsman can be left where they are or caught and put outside. The easiest way to capture a spider is to place a glass or plastic container over the spider, and then

A male black house spider helping to keep the insect population down in the house – here feeding upon a house fly.

TANYA LOOS

slide a thin piece of cardboard (such as a postcard or greeting card) underneath the animal, taking care not to damage their legs! Then, ideally with the help of someone to open the door, place both container and card outside and let the spider run away. For the really large spiders, a plastic sheet such as a flexible chopping block may be used. This method is very effective but does get the heart rate up a bit as you position the container over the spider on the wall. If a huntsman or other spider is trapped in the sink or bath, and you would prefer not to catch it, simply place a towel half-in and half-out of the sink or bath to form a ladder and allow the spider to climb out unaided.

- White-tailed spiders are best caught and placed outside using the above method.
- Funnel-web spiders and other mygalomorphs should be treated with extreme caution and moved outside. Always wear gloves when catching these spiders. The Australian Reptile Park has the country's only funnel-web antivenom program and they are always looking for live spider donations! Their website has useful advice on how to safely catch a funnel-web, and a list of spider drop off points around New South Wales.[6]

Sharing your house with spiders

The spiders generally like to keep to themselves. All we need to do is avoid inadvertently placing our hand in a spider's web, such as a redback spider, or accidentally disturbing a spider such as a white-tailed spider as it lays hidden in clothing, sheets or shoes. Here are some hints that may help you and your family members be spider aware:

- Wear gloves when carrying out clean-ups of sheds or verandahs.
- Check first before placing your hand in dark spaces such as mailboxes.
- Inspect any items that may be left outside before use, such as gumboots, bike helmets, winter gear, camping equipment and other potential spider habitat that hasn't been used for a while.
- Turn back your sheets before hopping into bed.
- Give any clothes that have been laying around in your floor pile a good shake before wearing; same applies to shoes and gardening gloves, especially those that haven't been worn for a long time.
- Funnel-webs can remain submerged without any ill effects for a very long time indeed, so do not assume that any large spider floating in a pet's water bowl or the pool is dead. Bring pet water bowls inside.
- Draught excluders fitted to the bottom of any external doors should exclude funnel-webs as they are strictly ground living.

References

1. Whyte R, Anderson G (2018) *A Field Guide to Spiders of Australia*. CSIRO Publishing, Melbourne.

2. Rayor S (2016) Hidden housemates: Australia's huge and hairy huntsman spiders. *The Conversation*, 6 April.

3. Australian Institute of Health and Welfare (2023) Contact with living things, <https://www.aihw.gov.au/reports/injury/contact-with-living-things>.

4. Mammola S, Malumbres-Olarte J, Arabesky V, Barrales-Alcalá DA, Barrion-Dupo AL, *et al.* (2022) The global spread of misinformation on spiders. *Current Biology* **32**(16), 871–873. doi:10.1016/j.cub.2022.07.026

5. Isbister GK, Gray MR (2003) White-tail spider bite: a prospective study of 130 definite bites by *Lampona* species. *The Medical Journal of Australia* **179**(4), 199–202. doi:10.5694/j.1326-5377.2003.tb05499.x

6. Australian Reptile Park (2023) Spider venom program, <https://www.reptilepark.com.au/animals-at-the-australian-reptile-park/venom-program/ spider-venom-program/>

This black rat is feeding on spilled seed from a bird feeding station.

Remove rats and mice without harming wildlife

BLACK RATS AND HOUSE MICE eat stored food and planted vegetables, keep us awake at night, create a pungent odour, chew electrical wires and transmit disease! Tackling the problem involves both habitat modification and direct killing – while making sure any control measures avoid harming owls, lizards, possums and snakes.

LOCATION

Australia-wide.

SEASON

All year round, with an autumn spike in cooler states when rats and mice seek warmth and food in houses as the temperature drops.

SPECIES

Black rats *Rattus rattus*, or brown rats *Rattus norvegicus* in coastal areas, and the house mouse *Mus musculus*.

PHOTO: DONNA POMEROY

Native rodents

Australia-wide there are several native rodent species that may be confused with black rats and house mice, which vary from state to state, as illustrated in Table 2. These range from the large (and aptly named) giant white-tailed rat to the very small western chestnut mouse.

Australian houses and gardens also provide a habitat for a suite of rodent-like carnivorous marsupials such as antechinus and brush-tailed phascogales. Even bandicoots may be mistaken for rats! All of these native species are protected under state wildlife laws. As our native mammal fauna varies so much from state to state, it is best to contact your local wildlife rescue organisation or state museum for advice and identification help.

If you have determined that your unwelcome animals are native rather than introduced rats or mice, then the challenge will be how to reduce their impact without harming the animals:

- You can prevent access to your house with carpentry, or by blocking up any gaps with expanding foam, wire mesh or steel wool.

Table 2. Mammal distribution matrix from *The Field Companion to the Mammals of Australia*.[1] Introduced mammals in bold.

	Qld	NSW	Vic	Tas	SA	WA	NT
Black rat							
Brown rat							
Bush rat							
Swamp rat							
Black-footed tree rat							
Giant white-tailed rat							
House mouse							
Western chestnut mouse							
Grassland melomys							
Fawn-footed melomys							
Southern brown bandicoot							
Long-nosed bandicoot							
Brush-tailed phascogale							
Agile antechinus							
Yellow-footed antechinus							

Giant white-tailed rats are a formidable native rodent of Far North Queensland. In the cooler months they like to shelter in warm car engines – a couple of moth balls under the bonnet may persuade them to move on.

PATRICK TOMKINS

- Protect your aviary or chickens with mesh over any gaps and a concrete floor (see the section on protecting chooks on pp. 96–102).
- If you have antechinus or phascogale coming inside, you can erect a suitable nest box in your backyard; this may lure them away from the house.

Risks

To wildlife: Rodenticides are used widely to treat rodent populations in both urban and agricultural settings. This entry focuses on rodent control around houses and gardens, with a particular emphasis on avoiding the use of second-generation anticoagulant rodenticides (SGARs).

SGARs kill the rodent after one dose and remain in the animal's tissue. Compounds such as brodifacoum, bromadiolone and difenacoum are the three most common SGARs in products that are widely sold, and persist in liver tissue after consumption for very long periods – and thus present the greatest threat of secondary poisoning in non-target wildlife of all anticoagulant rodenticides in use.[2] Other SGAR ingredients to look out for and avoid include flocoumafen and difethialone. In contrast, first-generation rodenticides work cumulatively, killing the rodent after multiple doses. The active ingredients have a relatively short half-life in liver tissue, reducing the incidence and severity of secondary exposure and bioaccumulation.

SGARs are found in the bodies of animals that prey on mice and rats directly. A study of southern boobook owls in an urban and peri-urban area in Western Australia found that 72.6% of owls found dead or seriously ill tested positive for anticoagulant rodenticides.[2] Day-flying birds of prey such as black-shouldered kites, nankeen kestrels and brown falcons, and generalist predators such as kookaburras, ravens and crows, are also threatened by SGARs, although more research is required to determine the extent.[3]

Rodenticide poison can enter the food chain in other insidious ways. Another study, prompted by the discovery of eight dead powerful owls in the Melbourne area in less than 1 year, found SGARs (particularly brodifacoum) in 83.3% of powerful owls tested (18 owls).[4] Powerful owls are arboreal hunters and prefer prey such as gliders and possums, or sometimes roosting birds such as galahs. They rarely eat rats and mice. Both brushtail possums and ringtail possums have been taken into care with rodenticide poisoning,[5] so the researchers concluded that the powerful owls were being poisoned by eating possums that had been eating rodent bait in roofs.

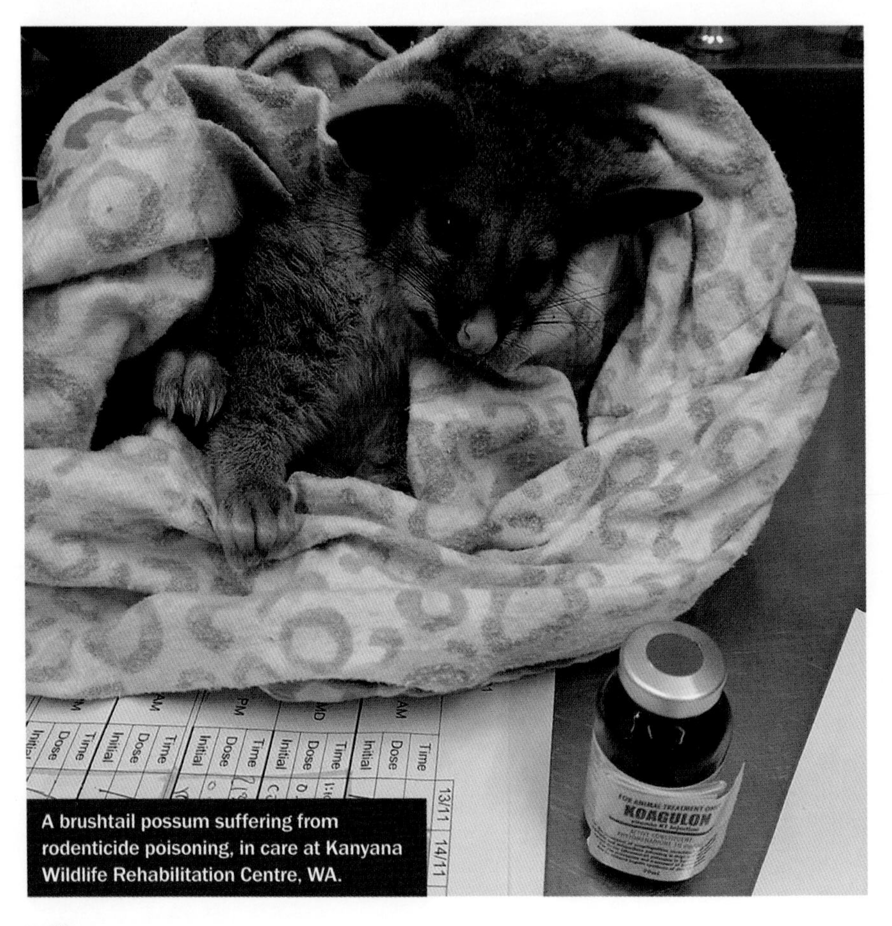

A brushtail possum suffering from rodenticide poisoning, in care at Kanyana Wildlife Rehabilitation Centre, WA.

KANYANA WILDLIFE REHABILITATION CENTRE

The owl-friendly rodent control campaign has been calling for the banning of SGARs in Australia to help our feathered predators, but lesser known is the fact that these compounds poison other animals. This includes those that eat rodents directly, such as dugites (a species of snake) in Western Australia, but also indirectly – bobtail lizards who may be eating at bait stations, and even tiger snakes exposed to SGARs. Tiger snakes eat frogs, not mice or rats, so researchers have concluded that rodenticides are well and truly in the food chain, at least in the Perth region where reptiles were tested.[6]

Rodenticide poisoning symptoms include weakness and disorientation, and many animals brought into care as a result of window or car strike may already be suffering the effects of rodenticide poisoning.

To people/property: Rodents are renowned disease carriers. Their gnawing at electrical wires causes house fires and machinery breakdown, including in cars. Rats and mice contaminate and destroy stored food, and cause crop wastage in both garden vegetables and fruit trees.

Actions and solutions

There are only two solutions that work. First, modify the habitat features enabling rats and mice to thrive in your house and garden and, second, reduce their numbers by killing the rodents directly using wildlife-friendly methods. Unfortunately, live capture and release is not recommended (see 'Is capture and release of rodents okay?'). Using a cat to control rodents is also problematic and unlikely to be successful (see 'Wildlife-friendly pet ownership' on pp. 148–156).

Habitat modification

Remove any open sources of food, both in the house and outside. Store all foodstuffs in glass jars and metal bins. Make sure any uneaten pet food is removed at night. Try to harvest all of your vegetable crop at once. Pick up fallen fruit in the garden. Hungry rats will feed on cat, dog and horse manure, so removal of this source of food may also help.

If you have an open compost heap, consider switching to a closed system. If you would like to keep your compost as an open heap, lay strong wire mesh at the base of the compost bay.

Rats and mice love nesting in piles of debris in forgotten corners of our gardens and sheds. Clean these up to remove these habitat areas. If you need to store wood or corrugated iron sheets, store them off the ground if possible. If rats are particularly persistent in a shed or under the house area, you may need to concrete the floor! This is an expensive solution, but in some cases the only way to remove their nesting and hiding habitats.

Remove plants such as ivy and palm trees from your garden, as these species provide habitat such as shelter for nesting and food such as berries and palm fruit.

Consider implementing a landscaping style that is not rodent friendly: break up existing dense plantings with exposed pathways, stretches of lawn or very low ground cover. Avoid growing ivy or other creepers from ground level up the side of your house – these are perfect black rat ladders! Leave a gap of at least 1 m between buildings and adjacent shrubs and trees.[7]

Reducing the rodent population

As animal lovers, deciding how to kill rodents can be difficult! In the past, poison may have been chosen as the mice and rats often die away from view after eating the bait, and no more handling of the animals is required. But in order to tackle a rat and mouse problem, we need to accept responsibility for the animal's welfare as well as minimise any harm to wildlife and pets.

The most humane way to kill a mammal such as a mouse or rat is blunt force trauma to the skull with a precision tool such as a hammer.[8] It is quick and painless. However, I don't know many people capable of carrying out this action with both the fortitude and skills required.

The next most humane option is a good quality snap trap.[8] Snap traps have a spring-loaded mechanism that releases when the animal takes the bait. You will have to buy different sizes depending on if you have a rat or mouse problem. Here are some tips for snap trap use:

- Choose traps with the widest opening – so they have as much force as possible when they snap shut.
- You will need five to 10 traps so you can trap an entire area simultaneously. Three traps set in a row ensures that a rodent trying to jump over the traps will be caught.[7]
- Before you lay traps, you will need to work out where the rats and mice are travelling safely from their nests to their feeding areas. Good spots are on the floor along walls, where you see droppings, and near entry and exit holes, behind furniture and other safe places are favoured.
- Rats are very wary of new objects in their environment – lay traps out with bait on them, without spring-loading them so the animals get used to eating the bait without harm.
- Good baits to use include peanut butter or other nut butter such as cashew, sultanas and oats. Only use a tiny bit – a pea-sized amount of bait will be enough to attract a rodent and set off the mechanism.

Many snap traps now available are cheaply made and have weak springs so are unlikely to kill humanely. What if the snap trap fails and the rat or mouse is still alive and needs to be killed? In the past I have personally drowned rats and mice that are caught in a trap but are still alive, hoping that that this was a humane solution. However, drowning takes time and is not considered euthanasia;[9] indeed the animal welfare laws related to the use of traps in Victoria state clearly that 'rodent kill traps must not be designed to drown an animal'.[10] Blunt force trauma to the head with a hammer is the least cruel method, as mentioned before, but it's a tricky thing to achieve in practice. The production of snap traps should be regulated and tested to identify those that cause rapid unconsciousness, and thus the least amount of trauma.[8]

Another problem with snap traps is that they are unselective and the baits used for rodents are also attractive to antechinus. Cage traps offer an alternative. Euthanasia can be done by emptying the trap occupant into a large, tough plastic bag, then swinging that over your shoulder and quickly down onto a hard surface. This is a more achievable (for many people) way of applying blunt force trauma and kills instantly. Cage traps also allow for identification of the animal while still alive, thus making sure it is not a native species. Some wildlife organisations urge people to always use cage traps, and to send images of the trapped animals to experts for identification.

Is capture and release of rodents okay?

Understandably, many people are keen to avoid killing animals, and use a trap that captures the rodents alive, with the animals being released either nearby or in bushland. While this solution may seem less cruel, mice are creatures of habit and suddenly being transported to an unknown location is extremely stressful for them, and is likely to result in their deaths. Killing mice is simply not possible for some people, but on balance live capture and release of mice is better for wildlife than using rodenticide. In the case of larger and more aggressive rats, research in Sydney has shown that black rats are not as reliant on humans as previously thought. Black rats readily occupy native forest in the urban–bushland boundary, far away from houses, preying upon nestling birds and other native fauna.[11] Catching and releasing rats into bushland is bad for the local wildlife, and so is not recommended.

References

1. Van Dyck S, Gynther I, Baker A (Eds) (2013) *The Field Companion to the Mammals of Australia*. New Holland Publishers, Sydney.
2. Lohr MT (2018) Anticoagulant rodenticide exposure in an Australian predatory bird increases with proximity to developed habitat. *The Science of the Total Environment* **643**, 134–144. doi:10.1016/j.scitotenv.2018.06.207
3. Debus S (2022) *Australian Falcons: Ecology Behaviour and Conservation*. CSIRO Publishing, Melbourne.
4. Cooke R, Whiteley P, Jin Y, Death C, Weston MA, *et al.* (2022) Widespread exposure of powerful owls to second-generation anticoagulant rodenticides in Australia spans an urban to agricultural and forest landscape. *The Science of the Total Environment* **819**, 153024. doi:10.1016/j.scitotenv.2022.153024
5. Wildlife Health Australia (2023) 'Rodenticide toxicity in Australian wildlife'. Canberra, <https://wildlifehealthaustralia.com.au/Portals/0/ResourceCentre/FactSheets/Multiple/Rodenticide_Toxicity_in_Australian_Wildlife.pdf>.
6. Lettoof DC, Lohr MT, Busetti F, Bateman PW, Davis RA (2020) Toxic time bombs: frequent detection of anticoagulant rodenticides in urban reptiles at multiple trophic levels. *The Science of the Total Environment* **724**, 138218. doi:10.1016/j.scitotenv.2020.138218
7. CSPC Quarterly (2013) Protecting raptors from rodenticides. *Common Sense Pest Control Quarterly* **27**(1–4) Special Issue 2011. <https://www.birc.org/raptors.htm>.
8. Baker S, Ayers M, Beausoleil NJ, Belmain SR, Berdoy M, *et al.* (2022) An assessment of animal welfare impacts in wild Norway rat (*Rattus norvegicus*) management. *Animal Welfare (South Mimms, England)* **31**(1), 51–68. doi:10.7120/09627286.31.1.005
9. Ludders JW, Schmidt RH, Dein FJ, Klein PN (1999) Drowning is not euthanasia. *Wildlife Society Bulletin* **27**, 666–670.
10. Animal Welfare Victoria (2023) *About the prevention of cruelty to animals legislation*. <https://agriculture.vic.gov.au/livestock-and-animals/animal-welfare-victoria/pocta-act-1986/about-the-prevention-of-cruelty-to-animals-legislation>.
11. Adams MWD, Grant LS, Kovacs TGL, Liang SQT, Norris N, *et al.* (2023) Commensal black rats *Rattus rattus* select wild vegetation over urbanised habitats. *Oikos* **2023**, e09671. doi:10.1111/oik.09671

Coastal carpet python, a welcome visitor in a shed on the Sunshine Coast.

Python in the roof

IN SOME PARTS OF AUSTRALIA, pythons live quietly in our sheds and roofs, providing a valuable rodent control service. Pythons pose little threat to humans if left alone, but they can be a danger to domestic animals. Sharing our place is possible if we ensure our pets are adequately protected.

LOCATION

Western, eastern and northern parts of Australia.

SEASON

Through most of their range, any time of year. Between Brisbane and Sydney, pythons are more likely to shelter in the roof over the cooler winter months.

SPECIES

The carpet python *Morelia spilota* is widely distributed across Australia with several distinct subspecies that vary in body length and colouration. Many people are familiar with the diamond python *Morelia spilota spilota* in Sydney.

PHOTO: ETHAN MANN

Behaviour

The first sign that a python has taken up residence in your roof could be the sound of the snake moving, like a roll of carpet being dragged across the floor. During breeding season male pythons engage in male-to-male combat. This may be mistaken for a pair preparing to mate, but instead it's a fight for dominance with muscular body wrestling and sometimes even biting. A pair of large males in the attic engaged in combat can make quite a racket!

Like all snakes, pythons shed their skins periodically as they grow. Even their eyes have scales, and these turn a milky white colour before the shedding process begins. Snakes are quite vulnerable while shedding, and a quiet roof space is ideal. Some roofs have several or even dozens shed skins in the attic. The shed skin is much longer than the actual length of the snake because there is skin between overlapping scales.

Pythons also leave their scats where they rest. Snake scats vary depending upon the kind of prey that they are eating at the time, but are usually a bit runny and slimy, elongated, and may have a smaller white patch of concentrated urea. These scats can be quite large from the larger individuals.

Most carpet pythons reported in the media are described as 4 or 5 m in length, but this is usually exaggerated by excited residents and journalists. Pythons will stretch their bodies as they travel from roof space to nearby shrubs, making them appear much larger in size. Out of the various carpet python subspecies, the coastal carpet python is the largest, and adult sizes in suburbia are usually 1.5 m but may reach 3 m in length.[1]

Carpet pythons are ambush predators. They wait concealed in a likely spot, which could be a corner of the roof, inside a shrub, in a tree, or next to roosting flying-foxes or a bird bath, and then pounce from a stationary position. Special organs that are heat sensing pits on the sides of their jaws help them detect their prey. Pythons will use the same spot for weeks or even months if it proves fruitful.

A study of python diets in Brisbane and Ipswich found that pythons change their prey as they grow. The younger, smaller pythons ate mostly lizards, then with the snake's growth the size of prey increased to rats and birds (including aviary birds such as cockatiels, budgies and quail), and the largest pythons graduated to poultry and cats (both feral and domestic).[2]

Risks

To pythons: Urban pythons are at risk from car strike, dog and cat attacks (especially juvenile pythons) – and both direct attacks and even short distance translocation by humans, both of which can be fatal.[3]

To people/property: Pythons are not venomous and rarely attack humans. They may retaliate when attacked or cornered. Their bite may be painful and will need medical attention. Pythons shed skin and scats in roof spaces and on balconies. As mentioned previously, they occasionally prey upon aviary birds such as lovebirds, cockatiels, princess parrots and budgies, chickens and their eggs, and domestic pets such as cats and small dogs.[2]

Does a python need to be removed from the roof?

Most people accept that the brushtail possum is part of the Aussie backyard as the possums live in our roofs and forage upon the rich resources available in our parks and gardens, so why not accept the carpet python, too? The carpet python preys upon the possums, and uses the same kind of roof spaces for shelter. In monsoonal regions such as Darwin, our gardens with their abundant, well-watered vegetation function as refugia against the climatic extremes of drought and rain.[4]

While our first instinct may be that the python needs to be removed and taken to more 'natural' surroundings, our suburban gardens are unconventional habitats that have been fully embraced by carpet pythons.

Carpet pythons have taken advantage of the abundant resources available in semi-rural and suburban habitats, and often occur in greater numbers in these areas compared to surrounding bushland. Densely planted gardens provide food, water and shelter, and warm, dark cavities in our house and shed roofs substitute for the tree hollows that they use in bushland areas.[2]

Researchers have discovered that snakes do not fare well after relocation from people's properties into nearby bushland. In fact, one study of dugites (a species of brown snake) which radio-tracked a small number of snakes after capture found all four snakes taken into bushland subsequently died, and only 50% of those moved 200 m survived.[5]

Pythons have home ranges, and snake catchers or wildlife rescue people have no way of knowing whether they are taking the pythons out of their home range. Males have larger home ranges than females and are likely to be on the move in spring as they search for females.[4] In this case, your visiting python may be on the move anyway.

Female pythons lay their eggs in a favoured nest site, and often this site is an open compost heap. The mother broods the eggs, keeping them at a stable temperature by shivering her muscles. Being moved away from the nest when the eggs still need care can be disastrous for the clutch and, of course, very stressful for the snake as she tries to travel from the release area back to her nest.[6] Relocated snakes need to travel more to find shelter and food, and this exposes them to predators such as cats and dogs, and makes them vulnerable to car strike.

Actions and solutions

Name your visitor

Snake advocates suggest naming your python helps reduce the natural fear and aversion we have for snakes. After all, who can be terrified of a creature called Monty?

Naming your python is also helpful when you can alert your neighbours: for example, 'Hey, if you see a large python, that is just Monty, he won't do any harm if you leave him alone, but just watch out if you own any small pets.' Acceptance of snakes is often a community exercise!

Times to leave Monty and friends alone include when they are digesting a large meal such as a possum (which can take weeks), or when shedding. They may lay in a bird bath, pond or even pool to soak the skin before shedding.

In the roof

If you are willing to put up with some noise, why not accept your guest as free rodent control? Make sure you stop using second-generation rodenticide as this will be very bad for your python (see pp. 49–56).

If you really do not want pythons in the roof, block up any small entry and exit holes after they leave – this is also helpful to discourage other roof visitors such as rats and possums.

In the house

If you have a resident python or two, chances are they may occasionally end up in the house. They could be curled up on the screen door, in your bathroom or, as one couple discovered, draped across the bed! If the python is in the house, you can open and close your internal doors as appropriate, and then open your external doors and windows and wait for them to leave – but as pythons may sit in one spot for days or weeks, this may not always be practical.

Call your wildlife rescue organisation first, and ask for a recommended snake catcher and, if possible, ask the catcher to release the python in the garden. Such a close release is the most humane for the python for the reasons described above, and usually the snake will not return. In Darwin, snake catchers are paid for by the Territory government, so snake callouts are free; this allows time for the catchers to educate people about living with snakes. This is an excellent model to reduce human-snake conflict as there is no cost for repeatedly taking the python friendly option ... and should be rolled out Australia-wide[7], creating snake consultants rather than snake catchers.

Screens on open windows can help discourage pythons from entering the house.

Protect your pets

Often when pythons consume prey such as guinea pigs, quail or domestic parrots, our small pets are actually the python's secondary choice. Their primary quarry is the rats and mice that are attracted to your pets' food. Reducing or minimising any rodent population will mean pythons are less likely to come near our pets (see pp. 49–56).

A very secure, well-maintained chicken coop or aviary is a must in python territory (see pp. 96–102).

Keep your cats inside and reduce dog interactions (see pp. 148–156), and keep your dog and cat food bowls inside.

If the worst happens and your cat or dog is discovered in a python's squeezing embrace, don't panic. Instead grab a spray bottle with methylated spirits, or alcohol such as gin or vodka and then spray it into the python's mouth. This should make the python let go. If your pet is still breathing, it may survive, and will need immediate veterinary attention.[1]

References

1. Watharow S (2013) *Living with Snakes and other Reptiles*. CSIRO Publishing, Melbourne.
2. Fearn S, Robinson B, Sambono J, Shine R (2001) Pythons in the pergola: the ecology of 'nuisance' carpet pythons from suburban habitats in south-eastern Queensland. *Wildlife Research* **28**(6), 573–579. doi:10.1071/WR00106
3. Shine R, Koenig J (2001) Snakes in the garden: an analysis of reptiles 'rescued' by community-based wildlife carers. *Biological Conservation* **102**(3), 271–283. doi:10.1016/S0006-3207(01)00102-1
4. Parkin T, Jolly CJ, de Laive A, von Takach B (2021) Snakes on an urban plain: temporal patterns of snake activity and human–snake conflict in Darwin, Australia. *Austral Ecology* **46**(3), 449–462. doi:10.1111/aec.12990
5. Wolfe AK, Fleming PA, Bateman PW (2018) Impacts of translocation on a large urban-adapted venomous snake. *Wildlife Research* **45**(4), 316–324. doi:10.1071/WR17166
6. Chantelle Derez, Brisbane Python Project Facebook page.
7. Cornelis J, Parkin T, Bateman P (2021) Killing them softly: a review on snake translocation and an Australian case study. *The Herpetological Journal* **31**, 118–131. doi:10.33256/31.3.118131

Part 2: Backyard

This sulphur-crested cockatoo may look old, but is likely to be just 1 or 2 years of age.

An 'old' cockatoo with feather loss and a deformed bill

BIRDS DO NOT AGE LIKE we do, so one that looks old is actually very unwell. Feather loss and deformities in the bill and feet could be due to a virus that causes beak and feather disease. Unfortunately, as the recommended treatment for affected birds is euthanasia, preventing transmission is the best approach.

LOCATION

Australia-wide.

SEASON

Any time of year, but can peak in late summer after the breeding season as young birds are more likely to become infected.[1]

SPECIES

All parrots, but particularly sulphur-crested cockatoos and lorikeets.

PHOTO: PIXABAY/VICKI NUNN

Psittacine beak and feather disease

Psittacine beak and feather disease (PBFD), named after the parrot family Psittacidae, is an often fatal disease of parrots, lorikeets and cockatoos.[2] In aviary birds, particularly budgerigars, the disease is known as French moult. It is caused by the beak and feather disease virus (BFDV) and is most commonly reported in sulphur-crested cockatoos and lorikeets, particularly rainbow lorikeets and scaly-breasted lorikeets, but some parrots seem less susceptible and rarely contract PBFD, for example cockatiels and eclectus parrots.[3]

The disease has been reported in Australian parrots for over 120 years, with the earliest known outbreak occurring in the late 1800s in the Adelaide Hills, decimating the local red-rumped parrot population.[4]

BFDV is usually transmitted between birds through feather dust, saliva and faeces and even from mother to egg. It is a generalist virus, meaning that it can infect many different host species and not just members of the parrot family: in recent years, it has been detected in rainbow bee-eaters and even one dead fledgling powerful owl. Genetic analysis of the virus in the owl suggested that it probably caught the disease by eating a PBFD-affected lorikeet.[5]

The virus has also been detected in Gouldian finch, zebra finch, southern boobook, barn owl, Australian magpie, brown goshawk, laughing kookaburra, Australian raven, tawny frogmouth, Australian white ibis and wedge-tailed eagle.[6]

Behaviour

A flock of sulphur-crested cockatoos can look simply immaculate, with a coat of perfect bright white feathers offset by their jaunty yellow crest. One of the secrets to their 'bleach-bright' appearance are their powder down feathers. These specialised feathers are found in a few bird groups such as herons, pigeons, doves and some, but not all, parrots. The tips of these feathers create a powdery substance that is used by the bird when preening to keep its feathers clean and dry, similar to a dry shampoo. Instead of moulting, these unusual feathers grow continuously.

A cockatoo with PBFD will develop abnormal feathers with each successive moult. As their special powder down feathers are lost or damaged, the bird becomes off-white or dirty looking and they will often lose their yellow crest. As the disease progresses, the skin lesions characteristic of the disease worsen, causing the bird to become bald and lose their ability to fly due to the loss of their flight feathers. In cockatoos the bill appears black and shiny in the absence of powder down. The bill and claws may start to grow abnormally.

In lorikeets, PBFD manifests mainly in the loss of the flight feathers; in New South Wales and Queensland this is so frequently reported that afflicted birds are known as 'runners'. Runners are most often rainbow lorikeets and scaly-breasted lorikeets that can still forage for food but cannot fly. Runners are often mistaken for baby birds and are most commonly reported in spring breeding season (see pp. 117–122 on baby birds).

Another indication of PBFD may be subtle colour changes in the feathers such as yellow flecks throughout green plumage in lorikeets. Gang-gang cockatoos may be suffering from PBFD for some time before any noticeable feather loss.[7]

Risks

To parrots: Even though it affects cockatoos and lorikeets quite differently, by the time the effects of PBFD are noticeable the disease can cause considerable suffering and eventually death in affected individuals. The crusty, 'old' cockatoo with a balding head, dirty and missing feathers is likely to be only a few months to 2 years old. It will develop secondary infections that often lead to death, as infection with the virus weakens the bird's immune system. Infected birds sustain internal tissue lesions that are not outwardly visible, but are potentially very painful for the bird. If their bill is misshapen and deformed, eating is difficult. Lorikeet 'runners' are very vulnerable to predation, plus it is very stressful for the bird, being unable to fly and forage effectively or keep up with their flock mates.

Beyond the obvious animal welfare considerations for individual parrots, the virus does not pose a significant threat to our more common species, but outbreaks of PBFD can be catastrophic in captive breeding programs for critically endangered species such as the orange-bellied parrot. An outbreak in the orange-bellied parrot breeding program in 1985 led to PBFD being listed in April 2001 as a key threatening process under the *Environment Protection and Biodiversity Conservation Act 1999* for at least 25 parrot species and subspecies. An outbreak of PBFD in these tiny populations can be disastrous due to the lowered immunity and increased stress experienced by very small populations.[8]

In parrots with large populations, such as cockatoos, rosellas and lorikeets, the birds have some immunity to the virus – some seem to recover (e.g. young crimson rosellas usually don't show any signs of PBFD), but it is not clear whether the virus stays hidden in their organs to re-emerge later. Some birds may remain infectious for months, presumably shedding the virus as they forage and rest in a flock with others.[1] Localised outbreaks can and do happen: for example, a small population of gang-gang cockatoos from the Hornsby and Ku-ring-gai local government area has a high prevalence of the disease.[9]

To people/property: None.

Actions and solutions

There is no known cure for a bird that is showing signs of PBFD, and euthanasia at a veterinary practice is the most humane option to prevent suffering in the bird, as well as further transmission to other local parrots.

Whether it is an unwell cockatoo or a runner, it is probably best to call your local wildlife rescue organisation for advice (BFDV is highly infectious and persistent, so let those you call know it may be a case of suspected PBFD). Catching a sick cocky that can still fly can be a long and tricky process. Once the cockatoo or lorikeet is captured, the right hygiene protocols should be followed. Wash your hands very well after handling an infected bird or any equipment that has been in contact with the bird. If you have companion birds or an aviary, make sure you keep the infected bird, as well as any equipment used to catch the bird, such as towels and gloves, separated from your own pet birds.

Prevention is best

PBFD is the greatest argument against feeding wild parrots. The crowding of birds at shared bird feeders and bird baths can facilitate transmission of the virus, as can contamination of food and water with bird faeces, feather dust and saliva. Ideally, the welfare implications of the disease, and the threats posed by PBFD to threatened parrots such as gang-gang cockatoos and swift parrots, will motivate those that love feeding seed to wild parrots to reconsider.

If you are feeding wild birds and you notice a sick cockatoo, stop feeding the parrots and/or remove your feeding station for a few weeks or even a couple of months. The BFDV is very persistent and will remain in hollows and nest boxes for at least three months.[1]

Keep bird baths scrupulously clean. If your bird bath has been used by an individual affected by PBFD, it is a good idea to remove the bath for a few weeks as a precaution.

Other tips

Many people, including wildlife carers, are tempted to keep PBFD-affected birds alive and look after them in the hope that they regrow their feathers, or that they live out their days feeling loved and cared for while in quarantine. Parrots are wonderful creatures, with big eyes and engaging personalities, and when they go bald they may also become even more cute and endearing. Remember that birds tend to hide that they are unwell as long as they possibly can, to protect themselves from predators. So while the bird in your care may look happy and content, it may be suffering considerably.

We need to consider whether we are keeping the affected parrot alive to protect our feelings, or for the welfare of the bird.

References

1. Martens JM, Stokes HS, Berg ML, Walder K, Bennett ATD (2020) Seasonal fluctuation of beak and feather disease virus (BFDV) infection in wild crimson rosellas (*Platycercus elegans*). *Scientific Reports* **10**(1), 7894. doi:10.1038/s41598-020-64631-y
2. Sarker S, Ghorashi S, Forwood J, Bent SJ, Peters A, *et al.* (2014) Phylogeny of beak and feather disease virus in cockatoos demonstrates host generalism and multiple-variant infections within Psittaciformes. *Virology* **460–461**(1), 72–82. doi:10.1016/j.virol.2014.04.021
3. Raidal SR, Peters A (2018) Psittacine beak and feather disease: ecology and implications for conservation. *Emu - Austral Ornithology* **118**(1), 80–93.
4. Ashby E (1907) Parakeets moulting. *Emu - Austral Ornithology* **6**, 193–194.
5. Sarker S, Lloyd C, Forwood J, Raidal SR (2016) Forensic genetic evidence of beak and feather disease virus infection in a Powerful Owl, *Ninox strenua*. *Emu - Austral Ornithology* **116**(1), 71–74.
6. Amery-Gale J, Marenda MS, Owens J, Eden PA, Browning GF, *et al.* (2017) A high prevalence of beak and feather disease virus in non-psittacine Australian birds. *Journal of Medical Microbiology* **66**(7), 1005–1013. doi:10.1099/jmm.0.000516
7. Peters A, Patterson EI, Baker BGB, Holdsworth M, Sarker S, *et al.* (2014) Evidence of Psittacine beak and feather disease virus spillover into wild critically endangered Orange-bellied parrots (*Neophema chrysogaster*). *Journal of Wildlife Diseases* **50**(2), 288–296. doi:10.7589/2013-05-121
8. Das S, Smith K, Sarker S, Peters A, Adriaanse K, *et al.* (2020) Repeat spillover of beak and feather disease virus into an endangered parrot highlights the risk associated with endemic pathogen loss in endangered species. *Journal of Wildlife Diseases* **56**(4), 896–906. doi:10.7589/2018-06-154
9. Department of Agriculture, Water and the Environment (2022) '*Callocephalum fimbriatum* (Gang-gang Cockatoo) conservation advice'. Department of Agriculture, Water and the Environment, Canberra.

Flying-foxes can suddenly arrive in great numbers when the right trees are in bloom. Here we see grey-headed and black flying-foxes.

Flying-fox backyard visitors

FLYING-FOXES USE A FLUID NETWORK of day roosts as they travel up and down the east coast. Long-distance fliers, the flying-fox visiting a garden in Sydney could have very well been in another state the week prior! Their pollination and seed dispersal activities link our increasingly fragmented forests and woodlands.

LOCATION

Northern Australia as far west as Broome, the east coast as far south west as approximately Geraldton, WA.

SEASON

Patterns of which camps are occupied, and how flying-foxes use the surrounding landscape (including gardens), are highly variable both between years and between seasons.

SPECIES

Grey-headed flying-foxes, black flying-foxes, spectacled flying-foxes and little red flying-foxes.

PHOTO: JUSTIN A WELBERGEN

Behaviour

Quite different from the generally smaller, insect-eating microbats, flying-foxes are the ultimate forest gardeners of Australia's rainforests and woodlands. Colloquially known as 'megabats' or 'fruit bats', flying-foxes leave their day camps each night to fan out across the landscape, searching for the night's meal of pollen, nectar and fruit. As they feed, the bats play a vital role in the pollination and seed dispersal of hundreds of species of native trees and shrubs – predominantly from the families Myrtaceae (*Eucalyptus, Callistemon, Angophora* and *Syncarpia*) and Proteaceae (*Banksia*). Flying-fox fur is thick and fluffy and thus very suited to capturing pollen. As the bats fly from tree to tree, or from one forest patch to another, they pollinate flowers like giant furry bees or honeyeaters.[1]

Flying-foxes are also effective seed dispersers of species such as lilly pilly and fig. Seed is spread via their droppings as they fly from feeding area to feeding area. This occurs over surprisingly large areas, as their nightly flights are often 20–50 km – one grey-headed flying-fox travelled 500 km in 48 hours![2]

Their diet explains why you may have lots of flying-foxes locally one week and then none the next, and why there may be great variability in flying-fox numbers generally. As the Australian landscape changes from season to season and from year to year, so too does the availability of flying-foxes' food. Camps across the country are occupied or empty depending upon local food sources. For example, many eucalypts

Grey-headed flying-fox feeding on pollen and nectar in a mass of blossom on a red bloodwood, *Corymbia gummifera*.

JUSTIN A WELBERGEN

flower only every 4 years or so. When a favourite food source, such as spotted gum, has a bumper flowering year in that region, flying-fox numbers will swell locally in their thousands as the bats eat the nectar and pollen from the flowers.

Camps are not only staging posts for the flying-fox's nomadic lifestyle – they are also nurseries! In the summer months, some camps are used as maternity colonies as flying-foxes mate and raise their young. Little red flying-foxes are 6 months out of sync with the other species, and they give birth in April–May. Flying-foxes breed slowly, raising just one pup per year or none, as some mothers may not breed every year.

After the young bat is born, the pup stays attached to the mother's nipple under her wing as she flies and forages. When the pup becomes too large to carry, they are left at the camp in a crèche area. The young will stay in the crèche each night as mum goes out to feed. Both mother and pup are incredibly vulnerable during this time. The energetic demands of lactation mean the mother needs adequate food each night, and if the worst happens to mum while she is out foraging, the pup will die.

Risks

To flying-foxes: Each flying-fox species is subject to a different range of threats, but something common to them all is the destruction of the high-quality habitat that supply the nectar, pollen and fruit the bats need. Flying-foxes are also very vulnerable to the effects of deadly heat waves, cyclones and extended rain periods, which can lead to food shortages and subsequently starvation.[3] The biggest risk flying-foxes face in gardens is entanglement in unsafe fruit tree netting and, on larger properties, barbed wire.[4] Power line electrocution is also common.[5]

To people/property: All bats, including flying-foxes, act as natural reservoirs for many kinds of diseases, often without showing symptoms or succumbing themselves. Indeed, many kinds of wild and domestic animals can be regarded as reservoirs of disease that may transfer to humans – for example parrots and psittacosis, cats and toxoplasmosis – and in bats, Australian bat lyssavirus (ABLV).

It is easy to avoid contracting ABLV: simply avoid touching flying-foxes (or any other bats). If you see a flying-fox in trouble, such as alone during the day or tangled in netting or on a fence, *always* call for help so that an experienced and vaccinated wildlife rescuer can assist. Even if you live right next to a busy flying-fox colony or have flying-foxes visiting your trees regularly, there is no risk of catching ABLV unless you are bitten directly, as ABLV is not spread through droppings or urine. As the saying goes: no touch, no risk.[6]

As well as ABLV, flying-foxes are hosts to the Hendra virus, which can spill over into horses. In rare cases, the resulting infection can result in the death of the horse and, in even rarer cases where a human has been in close contact with an infected horse, also humans.

There is no evidence of transmission of the Hendra virus directly from bats to humans. If you own horses, get them vaccinated against Hendra as a precautionary measure, and do not feed or water horses beneath trees where flying-foxes roost or visit regularly.[7]

The flying-fox friendly backyard

Flying-foxes were once very common across the east coast of Australia, with one single grey-headed flying-fox camp numbering in the millions.[8] Today, only the little red flying-fox has camps approaching these numbers. Grey-headed flying-foxes and spectacled flying-foxes are common garden visitors that are now listed as vulnerable and endangered respectively due to their ongoing population declines.[3]

Some people may feel like there are more flying-foxes than ever, but this is only because flying-foxes appear to be choosing to camp closer to humans, for reasons that are still unclear.[9,10]

A flying-fox friendly backyard is easy to achieve and may help reverse some of these precipitous declines:

- First and foremost – use wildlife friendly fruit netting! Fruit netting that has holes wide enough to poke your finger through is dangerous and means that a claw or wing can become entangled.
- Avoid the use of barbed wire (see pp. 157–162).
- Plant flying-fox friendly trees and shrubs – contact your local wildlife rescue organisation for advice on which species may be suitable.
- Remove cocos palms entirely – these plants are dangerous to flying-foxes, which are attracted to the palm fruit. Unfortunately, the fruit is poisonous when unripe, wears their teeth down and the large seeds can get stuck in their jaws! The flying-foxes can also get trapped and tangled in the palm fronds.
- Support positive action through your local council by educating the public about the beneficial aspects of flying-foxes, and any flying-fox camp management that is focused on keeping the bats where they are (see 'Flying-foxes and councils').

Actions and solutions

A busy flying-fox camp can generate plenty of noise, smell and droppings from bats – especially as numbers swell by the thousands when a local eucalypt is in mass flower. These effects are transient and tolerated well by those residents who are willing to live with some inconveniences as part of living in urban areas with bushland.[11] However, sometimes it can feel overwhelming to live close to large numbers of animals, and community members may be concerned about the associated noise, smell or disease risks.

The following advice is for those who like their bats, but are looking to mitigate some of the effects of living alongside them.

Sound and smell

A camp is a noisy affair, as flying-foxes are highly intelligent and social, and seem to have plenty to say to one another. Flying-foxes are described in one scientific article as having 'extremely loud vocalising in close proximity to one another' so they are definitely not using their 'inside voices'![12]

Flying-fox camps are louder at dusk and dawn as the camp arrives and leaves, and during breeding season. Lower frequency harsh chuckles and 'chups' are used during arguments over favoured branch roosts and during mating. The birth and raising of young pups also require a lot of communication. Young flying-foxes have high frequency 'isolation' calls, which help the mother bat find her pup when she returns from feeding each night.[13]

As well as their calls, like many other mammals, flying-foxes use their sense of smell to communicate with one another. The distinctive odour of a flying-fox colony is a musky scent that comes from their scent glands, not their droppings, and is used to help mark territories on branches and find mates.[14]

JUSTIN A WELBERGEN

Flying-foxes have a slow reproductive rate, and high infant mortality, making this grey-headed flying-fox pup all the more precious.

Flying-foxes and councils

Councils play a significant role in flying-fox management: a recent study of grey-headed flying-foxes recorded 546 known camp sites for grey-headed flying-foxes across 85 local government areas.[16] These councils need to balance the needs of the community while following the state and federal environmental laws that protect flying-foxes.

For well-resourced and environmentally minded councils, camp management minimises harm to the bats, keeping the camp in place rather than dispersing the flying-foxes. Meanwhile community education campaigns highlight the positive ecological role of the bats as keystone pollinators and seed dispersers. These councils may also provide environmental subsidies for those living very close to camps, such as funds for double-glazing or high-pressure hoses. In rural regional areas with a low ratepayer base, councils simply do not have the resources to manage a large influx of flying-foxes, particularly if the camp is recently established. In some local government areas, camps are dispersed, either by driving the bats away or by removing the trees the flying-foxes are roosting in. This approach is often taken due to anti-flying-fox councillors or negative local media coverage.[17]

Camp dispersal is very expensive, highly stressful for the flying-foxes and ultimately makes the problem worse, as splinter camps end up in surrounding areas and often in people's gardens. A recent review assessed 48 dispersal attempts from 1992 to 2020 and found that less than 25% of attempts could be considered successful.[17]

Part of the reason for the failure of dispersal is that the number of individual flying-foxes occupying a camp at any one time changes rapidly. In order to find out just how fluid the flying-fox population is, researchers attached satellite transmitters to 201 flying-foxes (including grey-headed flying-foxes, black flying-foxes and little red flying-foxes) for a period of 5 years. As long suspected, the study found that flying-fox camps are staging posts, with up to 17% daily colony turnover.[16] So the dispersal of one group of flying-foxes will be followed by arrival of new individuals from other parts of Australia that week, and the next week and the next!

Camp dispersal only works (to a point) at high profile, well-funded sites, such as the Melbourne and Sydney botanic gardens. The bats did not end up where intended after dispersal activities, and the ongoing long-term financial commitment of staff time, ecological expertise, site management and flying-fox dispersal follow-up for years is one that is beyond the reach of most councils.

Camps can be very noisy during breeding season, but noise levels will increase at any time of year if the camp is disturbed by dogs, lawn-mowers other loud noises. So keeping human activity quiet around camps actually reduces overall camp noise![15] Double-glazing of windows helps some residents who live very close to a camp.

Droppings

The droppings of nectar, pollen and fruit-eating animals are frequent, voluminous and very liquid in consistency, as they have such a short passage time through the gut. Flying-fox droppings are also accompanied by what are called 'spats' – little pellets of the chewed fibres of fruits where the juices have been consumed and the fibres pulped against the roof of the mouth, rather than eaten and taken into the digestive system.

To remove flying-fox droppings from your car, cover the stain with a wet cloth or newspaper and leave it to soak for about half an hour, and then simply wipe it away. Do not leave droppings on the car to 'bake' in the sun. A car cover during high periods of flying-fox activity is advised.

It is best to bring washing in from an outside clothesline before dusk to avoid flying-fox droppings soiling laundry. Alternatively, washing can be dried under covered areas.

Other tips

- Find out what plant species flying-foxes prefer and avoid planting these, or remove them – especially if they are just outside your window. Some councils subsidise or pay for the removal of cocos palms.[15]
- Create visual, sound or smell barriers with fencing or hedges using plants that do not produce edible fruit or nectar-rich flowers.
- If you collect rainwater for drinking, first-flush diverters are recommended to remove contaminants before clean water is collected in the tank – this is equally helpful for bird and possum droppings. Inlets and outlets on rainwater tanks should also be screened and the tank covered.

References

1. Southerton SG, Birt P, Porter J, Ford HA (2004) Review of gene movement by bats and birds and its potential significance for eucalypt plantation forestry. *Australian Forestry* **67**(1), 44–53. doi:10.1080/00049158.2004.10676205
2. Welbergen JA, Meade J, Field HE, Edson D, McMichael L, *et al.* (2020) Extreme mobility of the world's largest flying mammals creates key challenges for management and conservation. *BMC Biology* **18**, 101. doi:10.1186/s12915-020-00829-w

3. Westcott DA, Heersink DK, McKeown A, Caley P (2015) 'The status and trends of Australia's EPBC-listed flying-foxes'. CSIRO, Australia.
4. Camprasse ECM, Klapperstueck M, Cardilini APA (2023) Wildlife emergency response services data provide insights into human and non-human threats to wildlife and the response to those threats. *Diversity* **15**, 683. doi:10.3390/d15050683
5. Mo M, Roache M, Haering R, Kwok A (2021) Using wildlife carer records to identify patterns in flying-fox rescues: a case study in New South Wales, Australia. *Pacific Conservation Biology* **27**, 61–69. doi:10.1071/PC20031
6. Wildlife Health Australia (2023) 'Australian Bat Lyssavirus fact sheet: August 2023'. Canberra, <https://wildlifehealthaustralia.com.au/Portals/0/ResourceCentre/FactSheets/mammals/Australian_Bat_Lyssavirus.pdf>.
7. Wildlife Health Australia (2021) 'Hendra virus in flying-foxes in Australia fact sheet: October 2021'. Canberra, <https://wildlifehealthaustralia.com.au/Portals/0/ResourceCentre/FactSheets/Mammals/Hendra_virus_and_Australian_Wildlife.pdf>.
8. Ratcliffe F (1947) *Flying Fox and Drifting Sand: The Adventures of a Biologist in Australia.* Angus and Robertson, Sydney.
9. Tait J, Perotto-Baldivieso HL, McKeown A, Westcott DA (2014) Are flying-foxes coming to town? Urbanisation of the spectacled flying-fox (*Pteropus conspicillatus*) in Australia. *PLoS One* **9**(10), e109810. doi:10.1371/journal.pone.0109810
10. Timmiss LA, Martin JM, Murray NJ, Welbergen JA, Westcott D, *et al.* (2020) Threatened but not conserved: flying-fox roosting and foraging habitat in Australia. *Australian Journal of Zoology* **68**, 226–233. doi:10.1071/ZO20086
11. Larsen E, Beck M, Hartnell E (2002) Neighbours of Ku-ring-gai Flying-fox Reserve: community attitudes survey 2001. In *Managing the Grey-headed Flying-fox as a Threatened Species in New South Wales.* (Eds P Eby and D Lunney) pp. 225–239. Royal Zoological Society of New South Wales, Mosman.
12. Pearson T, Clarke JA (2019) Urban noise and grey-headed flying-fox vocalisations: evidence of the silentium effect. *Urban Ecosystems* **22**, 271–280. doi:10.1007/s11252-018-0814-2
13. Christesen LS, Nelson J (2000) Vocal communication in the grey-headed flying-fox *Pteropus poliocephalus* (Chiroptera: Pteropodidae). *Australian Zoologist* **31**(3), 447–457. doi:10.7882/AZ.2000.005
14. Hall L, Richards G (2000) *Flying Foxes: Fruit and Blossom Bats of Australia.* UNSW Press, Sydney.
15. Mo M, Roache M, Demers MA (2020) Reducing human-wildlife conflict through subsidizing mitigation equipment and services: helping communities living with the grey-headed flying-fox. *Human Dimensions of Wildlife* **25**(4), 387–397. doi:10.1080/10871209.2020.1735580
16. Currey K, Kendal D, Van der Ree R, Lentini PE (2018) Land manager perspectives on conflict mitigation strategies for urban flying-fox camps. *Diversity* **10**(2), 39. doi:10.3390/d10020039
17. Roberts BJ, Mo M, Roache M, Eby P (2020) Review of dispersal attempts at flying-fox camps in Australia. *Australian Journal of Zoology* **68**, 254–272. doi:10.1071/ZO20043

These young blue-tongue lizards were delivered by C-section after their mother was admitted to Australia Zoo Wildlife Hospital following a dog attack. Mum sadly didn't make it but the young lizards were successfully returned to the wild.

Caring for blue-tongue or bobtail lizards

BLUE-TONGUE LIZARDS ARE OFTEN CALLED blueys or simply blue-tongues. Shinglebacks have many names, including sleepy lizards, pinecone lizards, stumpytails and bobtails. Blue-tongues and bobtails are long-lived and loveable lizards who may occupy a garden for decades provided you follow a few tips to keep them safe and healthy.

LOCATION

Blue-tongues and/or bobtails of various species occur across most Australian cities.

SEASON

Spring and summer in southern states, all year round elsewhere.

SPECIES

Eastern blue-tongues occur in eastern Australia. Blotched blue-tongues occur as far south as Tasmania. The western blue-tongue is found in Western Australia. Bobtails occur in Western Australia, South Australia and parts of Victoria.

PHOTO: WILDLIFE WARRIORS

Behaviour

Several species of blue-tongue lizards may be found around Australia. The genus *Tiliqua* is characterised by their blue tongue and large size, thick body, large heads and short legs which create a boxy shape. The bobtail has these features plus distinctive, protective scales and very short and fat tails which are thought to confuse predators by being a similar size and shape to the head.

The most widespread species of blue-tongue is the eastern blue-tongue which occurs across the top half of Australia all the way down the east coast to Melbourne and Adelaide. In cooler regions such as Tasmania, the eastern blue-tongue is replaced by the blotched blue-tongue. One of the reasons these lizards do rather well in our backyards is their wide-ranging diet. They consume snails, slugs, insects, fungi and plant matter, as well as some more unconventional items. A study of blue-tongue diets in Sydney analysed the stomach contents of road-killed blue-tongues and found evidence of all the above items as well as pet food, newspaper, human hair, chicken bones and watermelon seeds.[1]

The blue-tongues you observe in your garden are likely to be residents. And they may have been at your place longer than you have as they can live for 30 years or more! Once a blue-tongue settles in their home range they tend to be very sedentary. The above-mentioned study radio-tracked 17 adults for 6 months, and found that 70% of their time was spent sheltering in two to four of their favourite shelter sites. The most movement recorded was during breeding season, when males are on the move seeking a mate.

Blue-tongues and bobtails are live-bearing – instead of laying eggs, they give birth. Blue-tongues can have litters of 10–15 young, which look like small versions of the adult. While their young are growing, the females move less, and the study found that the least movement was by pregnant females – in most cases pregnant females had home ranges of less than 1,000 m^2. One heavily pregnant blue-tongue was found sheltering deep within a compost heap.[1]

Blue-tongue lizards readily take to artificial or human made shelters. Old terracotta pots broken in half, concrete pipes, piles of bricks – the blue-tongues love them.

Blue-tongues are often mistaken for snakes, as sometimes just their tail is visible, protruding out of their favourite shelter – people see a large, scaled body part with stripes and their minds leap to 'snake'.

Bobtails are rarely mistaken for any other species as they are so distinctive. Perth is the only capital city with bobtails, but bush-block owners in South Australia and Victoria may also host these lizards in their backyards.

In South Australia, bobtails are called sleepy lizards, and a population there has been studied for decades by teams of researchers led by Professor Michael Bull.[2]

Back in the 1980s Prof. Bull noticed that each spring bobtails would pair up, and spend weeks together before mating.[3] Subsequent studies revealed that, unusually for lizards, bobtails are highly social with long-term, monogamous pair bonds. One couple has been recorded pairing up year after year for 27 years![2]

The bond is so strong that researchers have observed grieving behaviour. When a bobtail dies, for example by being run over by a car, the survivor will stay with their mate for days attempting to revive them and guard them from danger. This makes caring for bobtails in our backyards even more of a responsibility – your local bobtail could be someone's devoted partner!

Bobtails have a different reproductive strategy to blue-tongues. Rather than giving birth to dozens of young and then hoping some survive to adulthood, the female bobtail gives birth to two very large and well-developed young. When she is about to give birth, weight for weight it is the equivalent of a human woman giving birth to a 3-year-old child. The young will stay with the mother for months before moving off to find another home range.

Bobtails and blue-tongues are a real asset to the garden as they have a special fondness for garden snails; however, while blue-tongues are true omnivores, bobtails are mainly herbivorous, eating leaves, flowers, fruit and fungi.

Bobtails and blue-tongues are very charismatic, and when fed they often allow close contact with humans. However, feeding wildlife, including lizards, is not beneficial for the lizards themselves. Access to a near constant supply of rich human food including fruit and vegetables does not suit their energy requirements. These large skinks have slow-paced lifestyles and need to eat only every couple of days. Their eating patterns also vary depending on their location and the time of year. At the start of breeding season they need plenty of food, but in winter or cooler periods they stop eating altogether as they enter brumation. Brumation is the reptile version of

A blotched blue-tongue lizard feasts on a favourite food: the garden snail.

SIMON WATHAROW

mammalian torpor or hibernation, a period of inactivity where the skinks rest during the cooler months in safe shelters. During this time, bobtails will live off the energy stored in fat in their tails.

The risks associated with feeding wildlife are detailed later in this book, pp. 168–175. Increasing numbers of bobtails are coming into care at Perth wildlife hospitals as a result of overfeeding and subsequently obesity.

If you are very keen to provide a treat for your lizards, grow strawberries that they can forage themselves!

Risks

To lizards: Adult blue-tongues and bobtails are attacked and often killed by dogs. The young are preyed upon by cats. Machinery such as mowers and whipper-snippers can injure and kill lizards. Road strike occurs especially in the breeding season and then later when the young disperse.[4] In Perth, bobtails suffer from a disease known as bobtail flu (see'Bobtail flu').

To people/property: If provoked, blue-tongues and bobtails may bite, and then tend to hold on! If you need to move a blue-tongue or bobtail to a safer spot (i.e. if the lizard is in the way of your car on the driveway), gently wrap a towel around them or use a broom to sweep them into a cardboard box.

Actions and solutions

Lizard-friendly landscaping

A wildlife friendly native garden with locally occurring species can supply leaves, flowers and fruit directly to bobtails. Shrubs and tussock plants such as *Lomandra* provide shelter sites, while a large flat rock or paver could be an ideal basking spot for blue-tongues and bobtails to soak up the sunshine. Surround the basking site with dense cover to allow the lizards to duck out of sight the moment they feel unsafe. A shallow dish of water for drinking is also welcome, especially during hot periods.

Deep leaf litter and mulch provides an opportunity for the lizards to bury themselves, and is also great habitat for invertebrate prey such as earthworms and beetles.

Avoid using snail baits as poisoned snails will then be consumed by bobtails and blue-tongues – allow your garden lizards to be your snail and slug clean-up crew instead.

Be lizard aware

Some dogs find blue-tongues and bobtails extremely alluring. Instead of running away or hiding, the lizards simply face their attacker and protrude their large tongues. They may also inflate their bodies and hiss. The blue tongue display is an anti-predator response intended to startle a kookaburra or other attacker. One minute the kookaburra is facing a dull brown potential meal and then suddenly there is a flash of striking blue colour! Unfortunately, this anti-predator response is not all that effective on dogs and many blue-tongues and bobtails come into care with dog attack injuries, which are often fatal. Dog attacks are more frequent during breeding season when males are on the move.[4] They also occur when dogs dig the lizards out of their resting shelters, either when in brumation or when pregnant.

Many people ask if their blue-tongues need to be relocated because they have dogs. Relocation is stressful and if your garden is suitable habitat, the lizards may just return straight away. If possible, it is better to find solutions that allow both dogs and lizards to coexist in your garden.

The radio-tracking study mentioned previously found that blue-tongues were most active during the day when people tended to be at work, so if your dogs are outside while you are out, make sure your outside dog area isn't the same area as your lizard habitat areas. This includes provision of water, location of the compost heap and, of course, keep your pet food inside as both kinds of lizard will come right to the house to eat dog food.

Wildlife aversion training for your dog or dogs is recommended - snake aversion training may work for lizards as well. Use dense prickly native shrubs or fences to prevent dog access to some areas of your garden.

Bobtail flu

Are your bobtails looking thin or out of sorts? A respiratory disease syndrome known as bobtail flu affects bobtails, particularly in the Perth region, although it may be present in wild populations in other states.[5] If you see a bobtail that is weak or unresponsive; has a thin, flat tail, prominent hips and spine, weepy eyes and nose, sneezing, pale and frothy mouth; and is bloated or is tucking its legs under its body, it may be suffering from bobtail flu. This disease is very infectious and can be fatal if left untreated. The good news for Perth residents is that Kanyana Wildlife Rehabilitation Centre has been treating bobtails for years and now has an effective protocol for bobtail flu - and at present some 78% of infected individuals are returning to the wild - after an extended stay in hospital! If you think that maybe one of your bobtails is looking a little off, do not delay, call your local wildlife rescue straight away and arrange to bring the bobtail into care for treatment.[6]

A bobtail in care at Kanyana Wildlife Rehabilitation Centre. Flu-affected bobtails are nebulised twice daily for the first 21 days of their 28-day treatment program.

KANYANA WILDLIFE REHABILITATION CENTRE

Before starting mowing or whipper-snipping, scan the area for lizards, and either wait until they move on or gently move them to another part of the garden.

Similarly, in months when the lizards are active, check under your car before backing out of the driveway.

References

1. Koenig J, Shine R, Shea G (2001) The ecology of an Australian reptile icon: how do blue-tongued lizards (*Tiliqua scincoides*) survive in suburbia? *Wildlife Research* **28**(3), 215–227. doi:10.1071/WR00068
2. Bull CM (1988) Mate fidelity in an Australian lizard (*Trachydosaurus rugosus*). *Behavioral Ecology and Sociobiology* **23**, 45–49. doi:10.1007/BF00303057
3. Norval G, Gardner MG (2020) The natural history of the sleepy lizard (*Tiliqua rugosa*) (Gray, 1825) – Insight from chance observations and long-term research on a common Australian skink species. *Austral Ecology* **45**(4), 410–417. doi:10.1111/aec.12715
4. Koenig J, Shine R, Shea G (2002) The dangers of life in the city: patterns of activity, injury and mortality in suburban lizards (*Tiliqua scincoides*). *Journal of Herpetology* **36**, 62–68. doi:10.1670/0022-1511(2002)036[0062:TDOLIT]2.0.CO;2
5. O'Dea MA, Jackson B, Jackson C, Xavier P, Warren K (2016) Discovery and partial genomic characterisation of a novel Nidovirus associated with respiratory disease in wild Shingleback lizards (*Tiliqua rugosa*). *PLoS One* **11**(11), e0165209. doi:10.1371/journal.pone.0165209
6. C Jackson, Bobtail Coordinator, Kanyana Wildlife Rehabilitation Centre, Perth, personal communication, 20 September 2023.

A southern brown bandicoot pauses during a digging session.

Bandicoots digging in the lawn

MOST STATES HAVE A BANDICOOT (known as quenda in Western Australia) species or two visiting backyard lawns at night, leaving small conical digging pits as they dig for grubs and roots. Bandicoots are wonderful little ecosystem engineers – benefiting local bushland areas and backyards alike: even lawns!

LOCATION

Australia-wide.

SEASON

All year round.

SPECIES

Southern brown bandicoot, eastern barred bandicoot, quenda, long-nosed bandicoot, northern brown bandicoot.

PHOTO: HELEN GREENWOOD

Behaviour

Various bandicoot species occur in every state, with varying degrees of rarity. For example, in Victoria, southern brown bandicoots are only observed in a small area east of Melbourne (Cranbourne). The eastern barred bandicoot is regularly seen in Tasmanian gardens, but only occurs in fenced conservation enclosures in Victoria. Quenda are commonly seen in the Perth region. The most common garden bandicoot in Sydney is the long-nosed bandicoot. In Brisbane, Cairns and Darwin gardens, your most likely visitor is the northern brown bandicoot. Table 3 lists the species of bandicoots that may occur in backyards around Australia.

Bandicoots are omnivorous, eating invertebrates and small vertebrates, such as skinks and frogs, as well as fungi and plant material including berries, roots, and seeds. Many of these foods are foraged from above ground, but terrestrial food such as fungi, beetle grubs and roots must be dug up. Bandicoots dig their characteristic conical pits using their distinctive front feet with elongated claws. The resulting pits are perfectly snout-shaped.

You may have heard how beavers make dams and create ponds that benefit a whole suite of animals and plants. They are referred to as ecosystem engineers. While on a smaller scale, the digging pits of bandicoots also greatly benefit the natural environment. The conical pits, and the resulting spoils piled by the holes, help aerate and improve water absorption in the soil, speed up leaf litter decomposition and assist in seed germination.[2] These factors are very important in the Australian bush with our abundant leaf litter and often thin hydrophobic (water-repelling) soils. It has been estimated that a single quenda will displace around 4 tonnes of soil a year.[3] The foraging activities of bandicoots also help to disperse beneficial fungi[4] and the improved breakdown of leaf litter even reduces fire risk in urban bushland patches![5]

Risks

To bandicoots: Cats prey directly on bandicoots, and bandicoots also die from toxoplasmosis spread in cat droppings.[6] Dogs can attack bandicoots while they are sheltering in their nests. Some people trap or poison bandicoots, mistaking them for

Table 3. Bandicoots from state to state: mammal distribution matrix from *The Field Companion to the Mammals of Australia*.[1]

	Qld	NSW	Vic	Tas	SA	WA	NT
Northern brown bandicoot							
Southern brown bandicoot							
Quenda							
Eastern barred bandicoot							
Long-nosed bandicoot							

introduced rats.[7] Bandicoots are also vulnerable to predation by foxes, car strike and changes in their habitat from fires and urban development.

To people/property: Bandicoots may dig in lawns and may also eat garden vegetables.

In Queensland and New South Wales, some people are concerned that bandicoots in the garden means that their pets are at greater risk of attack from the paralysis tick. Paralysis ticks (*Ixodes holocyclus*) are generalist parasites that will use a wide range of hosts, including mammals, birds and even reptiles, such as blue-tongue lizards. However, there is no evidence to show that the presence of bandicoots is linked to higher risk of ticks for your cat or dog. Tick abundance is more likely to be linked with changes in the local environment, such as prevalence of exotic weeds and changing temperatures due to climate change.[8]

Actions and solutions

The first step is to make sure the guilty party is in fact a bandicoot or quenda. Rabbits also dig in lawns, but their digging pits are square edged, rather than curved. Visually, the differences between quenda and rats can be hard to spot if you are only getting a quick glimpse. Both mammals may be a similar shade of brown, so try looking at tail length and behaviour. Bandicoots have tails that are shorter than the body length, while black rats have tails that are longer than the length of their body. They also move quite differently, with quenda bouncing and hopping, and rats scurrying and running, without hopping.

While their conical digging pits look like damage, as described above the digging action aerates the soil and bandicoots often feed on the larva of beetles that feed on the lawn. So, this temporary lawn damage is in fact helping the lawn overall if you are willing to periodically sacrifice a perfect lawn. If you feel the holes are unsightly, you can simply backfill the digging spoils into the pit.

Bright lights and strongly smelling fertiliser pellets are also said to deter bandicoots, but these methods are untested.

In one study of community attitudes to bandicoots in Brisbane, many residents with bandicoots in their backyard were fine with bandicoot diggings in the lawn, but a little less happy to share their vegetables.[9] If you have bandicoots eating your vegetable crop, try:

- using raised garden beds
- building bandicoot fencing. Try snake mesh with gaps no larger than 20 mm. For best results, the foot of the mesh should be buried to a depth of at least 150 mm, and the fence should rise at least 500 mm above the ground. Paint the mesh black in a non-toxic paint if you would like your bandicoot fencing to blend into the garden visually.

The bandicoot-friendly garden

Don't feed bandicoots

Most people with bandicoots are delighted to host the little diggers, but unfortunately some people encourage them to visit by feeding them. The practice is popular – in one study near Perth of 65 households, over a third reported feeding their quenda.[7]

In Western Australia, quenda regularly come into care at Kanyana Wildlife Rehabilitation Centre with severe dental decay, sinus infection and malnutrition from eating food provided by well-intentioned bandicoot lovers. Tragically, the bandicoots have to be euthanised humanely as they are so sick! Feeding wildlife is only ever okay in specific situations as directed by your local wildlife rescue organisation or state government (see 'Is it okay to feed wildlife?' on pp. 168–175).

If you want a close look, instead erect a wildlife camera so you can capture images or video footage of your garden visitors.

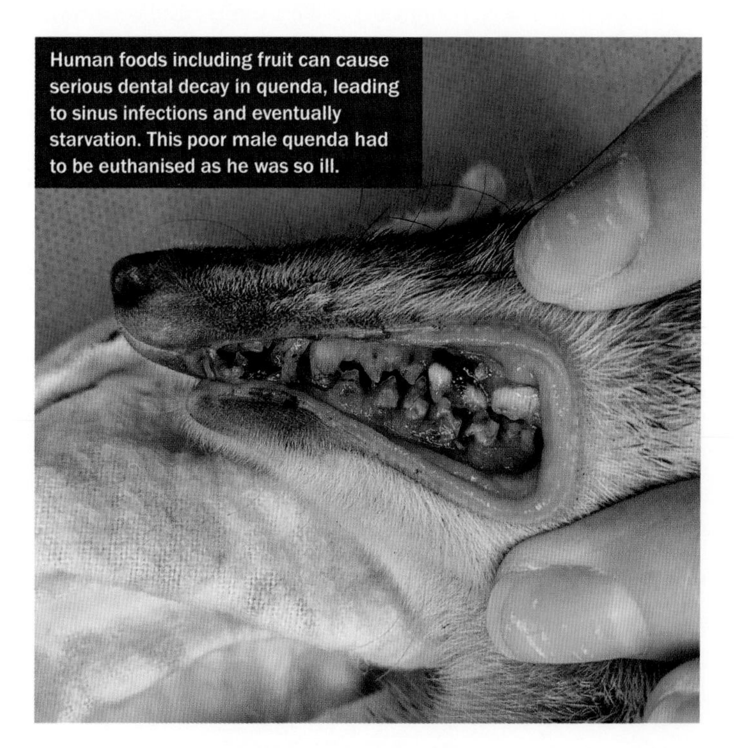

Human foods including fruit can cause serious dental decay in quenda, leading to sinus infections and eventually starvation. This poor male quenda had to be euthanised as he was so ill.

KANYANA WILDLIFE REHABILITATION CENTRE

Landscaping

Use native plantings that are masses of groundcovers. Continuous shrubs and native trees will allow the bandicoots to move around the garden without being exposed to predators. You can also create quenda corridors using terracotta pipes, hollow logs, rocks, wooden planks, bricks and roof tiles to link these habitats.

Be aware that this kind of landscaping also creates perfect habitat conditions for snakes, so incorporate landscaping principles that retain at least some visibility, such as gravel paths and a series of stepping stone habitats, rather than one bushy area (see the section on snakes in the garden on pp. 111–116).

Fencing

Garden fencing needs to be semi-permeable to ensure that bandicoots can enter and leave your garden, as it may form only part of their home range. Bandicoots can move through gaps at the bottom of a solid fence, but avoid the use of chicken wire on your boundary fencing. Better yet, use a hedge to mark your property instead of a fence.

Bandicoot safety

- Make sure your cat is an inside cat, thus protecting your bandicoots from both predation and disease.
- If your dog is not trained to avoid wildlife, fence off any bandicoot garden zones.
- Protect your lawn pest controllers by avoiding using pesticides or snail bait, especially those that use metalhyde, methiocarb or metsulfuron.
- If you own a pool, make sure there is a safety ramp extending from the skimmer box to enable any bandicoots who fall in to get out again. Cover the pool when not in use.

A bank of acacia hedging provides excellent cover for this garden bandicoot.

HELEN GREENWOOD

References

1. Van Dyck S, Gynther I, Baker A (Eds) (2013) *The Field Companion to the Mammals of Australia*. New Holland Publishers, Sydney.
2. Valentine LE, Bretz M, Ruthrof KX, Fisher R, Hardy GES, et al (2017) Scratching beneath the surface: Bandicoot bioturbation contributes to ecosystem processes. *Austral Ecology* **42**(3), 265–276. doi:10.1111/aec.12428
3. Ryan CM, Hobbs RJ, Valentine RE (2020) Bioturbation by a reintroduced digging mammal educes fuel loads in an urban reserve. *Ecological Applications* **30**(2), e02018. doi:10.1002/eap.2018
4. Hopkins AJM, Tay NE, Bryant GL, Ruthrof KX, Valentine LE, *et al.* (2021) Urban remnant size alters fungal functional groups dispersed by a digging mammal. *Biodiversity and Conservation* **30**, 3983–4003. doi:10.1007/s10531-021-02287-4
5. Ryan CM, Hobbs RJ, Valentine LE (2020) Bioturbation by a reintroduced digging mammal reduces fuel loads in an urban reserve. *Ecological Applications* **30**(2), e02018. doi:10.1002/eap.2018
6. Wildlife Health Australia (2019) 'Toxoplasmosis of Australian mammals'. Canberra, <https://wildlifehealthaustralia.com.au/Portals/0/ResourceCentre/FactSheets/Mammals/Toxoplasmosis_of_Australian_Mammals.pdf>.
7. Kristancic AR, Kuehs J, Richardson BB, Baudains C, Hardy GES, *et al.* (2022) Biodiversity conservation in urban gardens – pets and garden design influence activity of a vulnerable digging mammal. *Landscape and Urban Planning* **225**, 104464. doi:10.1016/j.landurbplan.2022.104464
8. Lydecker H, Stanfield E, Lo N, Hochuli D, Banks P (2015) Are urban bandicoots solely to blame for tick concerns? *Australian Zoologist* **37**(3), 288–293. doi:10.7882/AZ.2015.008
9. FitzGibbon SI, Jones DN (2006) A community-based wildlife survey: the knowledge and attitudes of residents of suburban Brisbane, with a focus on bandicoots. *Wildlife Research* **33**, 233–241. doi:10.1071/WR04029

A large male eastern grey kangaroo with powerful, muscular forearms.

Kangaroos in the backyard

IN SOME REGIONS KANGAROOS ARE an everyday part of people's lives, with these large and graceful animals encountered at golf courses, schools, retirement communities and, of course, in our backyards. Kangaroos make wonderful visitors, but their size and sometimes high numbers can rightly make people nervous!

LOCATION

Eastern grey kangaroos are found along the east coast, and in Tasmania where they are called forester kangaroos. The western grey kangaroo occasionally visits gardens in Perth and Adelaide.

SEASON

Any time of year, but kangaroos will come closer to houses during extreme weather periods, such as during heavy rain to seek shelter or in periods of drought in search of water (and watered lawns).

SPECIES

Eastern and western grey kangaroos.

PHOTO: PIXABAY/PEN_ASH

Behaviour

The eastern grey kangaroo and western grey kangaroo are closely related, with similar diets and lifestyles. Eastern greys are generally greyer in colour, and western greys are a rich chocolate brown – although colours do vary among individuals. The eastern grey kangaroo comes into contact with humans the most out of the two species,[1] and so will be featured for this entry but the advice is applicable for both kangaroos, and larger wallaby species.

Eastern grey kangaroos are sexually dimorphic, meaning that the males and females are quite different in size. Male kangaroos average around 70 kg in weight. Female kangaroos are a mere 35 kg, and much shorter in stature with fine, rather pretty features. The males grow throughout their lives and may reach 90 kg and up to 2 m tall.[2]

Kangaroos are highly social animals with complex social interactions including hierarchies and friendships between females. A dominant male (i.e. the 'head' male) has a group of females, and this position in the hierarchy is reached by fighting. Kangaroo sparring is like upright wrestling or boxing while balancing on their muscular tail – the hind legs are used less often than the arms. This is why the biggest males have such 'ripped' chests and massive biceps and triceps.

Eastern greys have a fission fusion society structure – consisting of many small groups that occasionally gather in larger groups called mobs – such as the 70 plus individuals gathered at a golf course. While they graze together out in the open, they are rarely more than 150 m from cover.[2]

Eastern greys are grazers of grasses and herbs, although keen gardeners and people working in revegetation will know that kangaroos also browse other plants such as shrubs.

Like the Australian magpie, the eastern grey kangaroo has benefited greatly from our love of lawns. We provide well-watered expanses of lush pasture, as well as water to drink and places to rest and hide in the form of shrubs and trees. This applies to golf courses and sports grounds – but also to many rural, semi-rural and even town properties.[3] And it seems the kangaroos do not mind being near houses at all.

Researchers wanted to know more about the behaviour of kangaroos in a large, semi-rural cluster of properties near Coffs Harbour, NSW. A combination of radio-tracking and GPS collars were attached to 14 male kangaroos – and over the course of the short study the tracking revealed that half their time was spent within 50 m of houses! This is remarkable as most of the 180 properties at Heritage Park are 2 ha or more in size.[4]

Risks

To kangaroos: While kangaroos have greatly benefited from resources available in human-dominated areas, the risks are great, and include wildlife-vehicle collision, dog attack, fence entanglement and diseases. In peri-urban areas that are fairly static, for example the edges of regional centres, kangaroos and humans can coexist reasonably well. In areas that are in rapid transition from rural or undeveloped land (kangaroo habitat) to housing developments kangaroos may become trapped within islands of remaining open land and may starve from declining food availability. In this situation these listed effects can be much worse and cause serious welfare concerns for the trapped kangaroos.[1]

To property/humans: There are only two recorded fatalities from kangaroos – one man was killed by a hand-raised 'pet' that became aggressive, and another man died trying to protect his dog. Kangaroo attacks may occur wherever there are large numbers of kangaroos in close proximity to humans. An analysis of kangaroo-human conflict in Coffs Harbour, NSW, found records of some 40 attacks and incidents over a 10-year period. Surprisingly it is not just the large males involved in these incidents: female kangaroos may attack a human if they feel their joey is unsafe.[5]

If provoked, kangaroos will retaliate and harm or kill dogs. Many so-called attacks, particularly where a dog is involved, are the result of the kangaroo trying to defend itself from a natural enemy (dogs/dingoes). Note that it is illegal to permit your dog to chase wildlife, including kangaroos, even if you think the dog is chasing 'just for fun'.

Kangaroos have very regular and well-defined routes from their feeding areas to resting areas each day. If fences are in the way, they can cause damage to fences over time.

Actions and solutions

In most cases, living safely with kangaroos simply requires some information on human-kangaroo etiquette. At a retirement village at Port Macquarie, out of 138 survey respondents, some 30% of residents sometimes felt unsafe leaving their houses. Despite this high figure, lethal control to reduce kangaroo numbers was unanimously opposed, and instead 92% felt that information about living with kangaroos should be provided to residents.[6]

In some cases, practical solutions such as fencing and landscaping may also help.

Human–kangaroo etiquette

Keep a safe distance from the kangaroos while you are going about your usual garden business. Perhaps if you see a large male and a couple of females between you and the letterbox or the clothesline, either go a different route if you have a larger property or wait until they have moved on.

Do not surprise them – you can talk to them, as this lets the kangaroo or kangaroos know where you are. Kangaroos have very regular home ranges and the individuals you see regularly are likely the same mob.[4] Many folks name their local kangaroos and therefore one can calmly say something like 'G'day Kevin I am just walking past you ...'.

This kind of interspecies connection and respect is easy to foster, as kangaroos will respond with different behaviours depending on landholder attitudes, with more kangaroos grazing on land that is kangaroo-friendly.[7]

Much like the southern cassowary (see pp. 130–135), there are particular behaviours and situations to look out for in your visiting kangaroos. If you see these behaviours, stay well away from the kangaroos until they settle down or move on.

Behaviours to be wary of include:

- sparring or fighting behaviour between males, which include dominance displays such as standing up very tall, walking slowly on all fours with back arched, rubbing chest on grass or grabbing grass tussocks with front paws

A female eastern grey kangaroo with her joey – at this age the joeys are referred to as 'young-at-foot', but they will still cram into mother's pouch if danger appears.

PIXABAY/SHARKALOT

- courtship or mating behaviour – if a male is touching or sniffing female kangaroos
- mother kangaroo behaviour – do not approach or move between a female kangaroo and her joey. Female kangaroos may attack if they perceive a threat to their young.

To keep the situation in your backyard respectful and calm, never feed your kangaroos. Feeding kangaroos reduces their natural fear of people and can make them very pushy and demanding, and they will attack humans for food, much like cassowaries. Increasing reports of kangaroo human conflict on the Fraser Coast, Qld, may be due to kangaroos being fed.[8]

If you feel threatened by a kangaroo:

- Never try to shoo them away. Avoid yelling, standing tall, waving your arms or throwing items at the kangaroo, as the kangaroo may interpret this as a challenge by a rival male kangaroo and attack in response.
- Do not turn your back to the animal and run, but instead give a short deep cough (the sound of a submissive kangaroo), avert your eyes and back away slowly while trying to make yourself appear smaller.
- Try to position a tree or fence between you and the animal.
- If the worst happens and the kangaroo starts to attack, and if you happen to have an object with you when the kangaroo approaches (e.g. a branch, stick or pole), hold it out in front of you so it acts as a barrier. This may deter an aggressive kangaroo – but never use the object to strike the animal.
- If the kangaroo does attack, drop to the ground, curl into a ball (if you are able to) or otherwise lie face down and use your arms to protect your head and neck. Try to remain calm and still until the animal moves away.[9]

If you have vulnerable family members

If you live in an area with abundant kangaroos and you are worried about older family members, young children or small dogs, the best thing to do is create a safe space near the house. The safe space will be surrounded by kangaroo barrier fencing that excludes kangaroos, which is 1.5 m high. This small area will need to have a gate to release any animals that *do* make their way in.

A smaller safe space is a better option than barrier fencing around the whole block. It is much less expensive as it requires less fencing material but, importantly, perimeter fencing may funnel the kangaroo problem into other areas.[1,5]

If fencing is not feasible, strategic landscaping such as tall, dense hedges can work to keep kangaroos away from areas near the house.

An old man 'roo comes to die

A large male kangaroo may be passing through, but if he hangs around for weeks, he may have been ousted from the kangaroo group or mob by the new dominant male. Houses in the bush or on the farmland/bush edge are usually at the outer edges of most of the group's home ranges, and provide resources such as fresh water, grasses and protection from the elements.

Your visitor may be recovering from bruises and perhaps wounded pride from his most recent battle. He may need 3 or 4 days of complete quiet at first. During this time, keep children and pets away from the kangaroo. Call your local wildlife rescue organisation if he seems injured (moves very reluctantly or is limping) or is unable to feed.

A few years ago, we had one of these visitors. After a couple of weeks of watching a large but rather weathered old male feeding on grass in various parts of the garden, our old man kangaroo passed away peacefully one night by our back door, under a blackwood tree. His teeth were very worn; it was such a privilege to have a wild animal feel safe to die a gentle death of old age in our space.

Tip: To move a large dead kangaroo, lay a blanket down, roll the animal onto the blanket and use three people to hoist the body into a wheelbarrow.

You can also deter kangaroos from visiting your property by reducing the amount of grass available, by keeping any lawns mown short, and considering the use of more garden beds and rockeries instead of lawn areas. Another option may be to clap your hands, bang short planks together, or bash pots and pans together to discourage the kangaroos from being too accustomed to close proximity with humans – from a safe distance, of course!

References

1. Herbert CA, Snape MA, Wimpenny CE, Coulson G (2021) Kangaroos in peri-urban areas: a fool's paradise? *Ecological Management & Restoration* **22**(1), 167–175. doi:10.1111/emr.12487
2. Richardson K (2012) *Australia's Amazing Kangaroos: Their Conservation, Unique Biology and Coexistence with Humans*. CSIRO Publishing, Collingwood.
3. Descovich K, Tribe A, McDonald IJ, Phillips CJC (2016) The eastern grey kangaroo: current management and future directions. *Wildlife Research* **43**, 576–589. doi:10.1071/WR16027
4. Henderson T, Vernes K, Körtner G, Rajaratnam R (2018) Using GPS technology to understand spatial and temporal activity of kangaroos in a peri-urban environment. *Animals* **8**, 97. doi:10.3390/ani8060097
5. WIRES, NPWS and CHCC (2016) 'Kangaroo management plan for the Coffs Harbour northern beaches'. Wildlife Information, Rescue and Education, National Parks and Wildlife Service, and Coffs Harbour City Council, Coffs Harbour.

6. Ballard G (2008) Peri-urban kangaroos. Wanted? Dead or alive? In *Too Close for Comfort: Issues in Human–wildlife Encounters*. (Eds D Lunney, A Munn and W Meikle) pp. 49–51. Royal Zoological Society of New South Wales, Sydney.
7. Austin CM, Ramp D (2019) Behavioural plasticity by Eastern Grey Kangaroos in response to human behaviour. *Animals (Basel)* **9**(5), 244. doi:10.3390/ani9050244
8. Bolton M, Jacques O (2022) Kangaroo attack on Fraser Coast leaves 67 year old Sunshine Coast woman with broken leg. ABC Sunshine Coast. < https://www.abc.net.au/news/2022-07-29/kangaroo-attack-leaves-67-year-old-woman-with-broken-leg/101282390. >
9. Queensland Government (2023) *Living with Kangaroos.* <https://www.qld.gov.au/environment/plants-animals/animals/living-with/kangaroos>

A spotted-tailed quoll feeds on a wallaby.

Protecting your chickens from quolls and other predators

KEEPING YOUR CHICKENS SAFE IS all about good husbandry and making sure you have the right housing. Although these measures involve an initial outlay of resources and may take a bit of building know-how, effective barriers are the most wildlife friendly way to protect your chickens.

LOCATION

Australia-wide.

SEASON

All year round.

SPECIES

In Tasmania, chickens will need protection from the spotted-tailed quoll, eastern quoll and Tasmanian devil. On mainland Australia, the introduced red fox is the main predator, as well as a range of carnivorous marsupials, birds of prey and reptiles.

PHOTO: ANA GRACANIN

Behaviour

For many people, chickens are also pets and part of the family, so attack by predators is more than just the loss of a source of eggs. The sound of frightened chickens at night may also be distressing. Chickens who are not protected from predators may become terrified and stop laying.

A flock of backyard hens represents a relatively easy meal for a large range of predators. In Tasmania, the spotted-tailed quoll, eastern quoll and Tasmanian devil will prey on your hens and their eggs if able to do so. In south-eastern Australia, foxes are the greatest threat with the occasional spotted-tailed quoll and brush-tailed phascogale attack. Heading north, foxes, large monitor lizards and pythons will access the hen house if they can; there's also a slight chance of spotted-tailed quoll and northern quoll attacks. Birds of prey such as grey goshawks, brown goshawks and sparrowhawks may attack hens as they free-range around the backyard, Australia-wide.

As there has been a rather surprising case of a quoll in a hen house in recent times, this entry focuses on protecting your hens from quolls. Chicken husbandry is a vast and varied subject and this entry touches on effective hen husbandry that is wildlife friendly, with the focus on quolls shining a light onto a source of potential conflict that still occurs to this day![1] Due to their taste for hens, in the past quolls were shot, trapped and poisoned in large numbers wherever they occurred. These days quolls are a threatened species, and it is illegal to trap, let alone kill, a quoll. Housing your hens safely from quolls works equally well for foxes and all the other predators listed here.

Spotted-tailed quolls are also called tiger quolls, despite the fact they are covered in white spots instead of stripes. But the descriptor 'tiger' is appropriate, as these cat-sized marsupials are the top native predator on the mainland. They are hypercarnivores, or meat specialists. Another hypercarnivore, the Tasmanian devil, used to occur across Australia but, as the name suggests, is now restricted to Tasmania. Both species are important scavengers as well as predators and a vital but sadly declining part of our woodland and forest ecosystems.

The spotted-tailed quolls' favourite prey is arboreal mammals such as greater gliders and possums, and one study of quoll scats (or faeces) in New South Wales revealed just how skilled these animals are at climbing. Quolls also hunt on the ground, where they prey on rabbits, bandicoots and pademelons. The study revealed that they are opportunistic hunters who vary their diet according to the changing availability of prey throughout the year. In winter, they eat mainly mammals whereas in the summer months, quolls feast on abundant insects, reptiles and, to a lesser extent, birds.[2]

Quolls spotted for the first time in centuries

With a very large home range and a secretive nocturnal lifestyle, you never know when you might spot a spotted tailed-quoll. The species was previously presumed extinct in South Australia, with the last official records in Mount Burr Forest in the 1880s. Locals said they had seen individuals in the 1970s and '80s, but these were unconfirmed, verbal reports. Then, in 2023, Pao Ling Tsai, a farmer in Beachport, trapped a spotted-tail quoll who had been preying upon his chickens – the first sighting in 130 years! The quoll was checked by a wildlife vet, treated for mange and released into an 'undisclosed area' of forest.

A similar story happened in the Grampians (Gariwerd) in 2013: park ranger Dave Roberts was stunned to see a spotted-tailed quoll captured on wildlife camera footage. After checking that the nearby Halls Gap Zoo had not lost a captive quoll, the record was confirmed as the first sighting since 1872! That is 141 years between sightings.

Although dense and remote woodland and forest are the main habitats of quolls, they do cross town and agricultural areas as they move around their very large home ranges, often turning up in some surprising places! See 'Quolls spotted for the first time in centuries'.

Risks

To wildlife: Quolls and their smaller relatives, the carnivorous marsupials known as brush-tailed phascogales, may become trapped in chicken coops, causing stress and injury to the trapped animal – and to your chicken flock. Young spotted-tailed quolls and blue-tongue lizards can become trapped as they squeeze through chicken wire. Goshawks and other birds of prey can enter poorly secured coops and then become trapped as they are unable to find their way out again.

To people/property: Some predators can damage coops if the mesh walls are not securely installed. If your hen house has a rodent population, this can attract venomous snakes, which can make cleaning out the chook house or collecting eggs a risky exercise for the landholder.

Actions and solutions

Prevention is best. Seek advice from permaculture or gardening groups in your area or local wildlife rescue organisation to work out what kind of predators, non-native and native, are likely to attempt to prey upon your flock.

Table 4. Predators of chickens and aviary birds. Introduced mammals in bold.

	Qld	NSW	Vic	Tas	SA	WA	NT
Red fox							
Tasmanian devil							
Spotted-tailed quoll					?		
Eastern quoll							
Brush-tailed phascogale							
Grey goshawk							
Brown goshawk							
Carpet python							
Olive python							
Monitor lizard							

Table 4 lists a range of predators that may prey upon chickens and aviary birds. Some of these animals, such as Tasmanian devils, foxes and quolls, look for any weaknesses in the chicken coop and force their way in. They may climb over tall fences or dig their way into the coop. Birds of prey such as goshawks are ambush hunters, and will swoop down from a protected perch and take free-ranging chickens and even enter uncovered coops.

The good news is that if you have a well-made coop, it is likely to keep your chickens safe from the largest to the smallest of predators. If you live in an area that has foxes, Tasmanian devils, quolls, goshawks or large pythons, it is a good idea to consider a permanent covered enclosure for your hens during the day.

Other tips

Make sure your hen house and run are clean and rodent-free

While hens and their eggs are the biggest drawcard, the presence of rats or mice may also attract predators; rodents are a source of food for pythons and brown snakes. Good hen-house husbandry to prevent rodents includes keeping the coops very clean by removing uneaten food scraps, feeding stations that are raised off the ground or foot-operated, and the use of quality materials and solid construction throughout. For more on wildlife-friendly rodent control, see pp. 49–56.

Say goodbye to 'chicken wire'

Use sturdy steel mesh instead. The wire needs to be too thick for a devil, quoll or fox to bite through, with spacing too narrow for a rodent. Use 10–15 mm^2 spot-welded aviary mesh. Another reason to avoid chicken wire is that the aperture shape and size can be deadly to juvenile quolls attempting to force their way in, as they may become stuck and strangled. Blue-tongue lizards can also get their head or sometimes their whole body stuck in the mesh. One year we found a huge blue-tongue who had died while stuck in our chicken wire vegetable garden fence. Their head and front legs managed to get through but then they got stuck halfway. If your coop chicken wire is rust free and sturdy, you can retrofit it to make it safe for lizards by adding a layer of smaller aviary mesh along the bottom at lizard height.

Protect your hens from below – and above

Place your chicken coop on a concrete floor for maximum impenetrability. If your coop has an earth floor, extend the mesh or walls at least 400 mm below ground. Cover your coop with a roof and or cage mesh canopy that has no gaps larger than 20 mm.

A hen aviary for safe daytime activity

In some areas, hens need to be protected during the day. Build your hens a secure enclosure that is as large as possible, allowing your hens to carry out all their natural behaviours – exploring flower beds and shrubs, dust bathing, with areas of shade and sun, and places to run, hide and explore. Don't forget the protective skirt of wire that is at least 400 mm deep.

Deter predators

Sensor lights can be an additional predator deterrent – the light comes on as the quoll or fox approaches the coop and startles it. Electric fences can be used as a short-term option to protect your flock if you need to rebuild your coop.

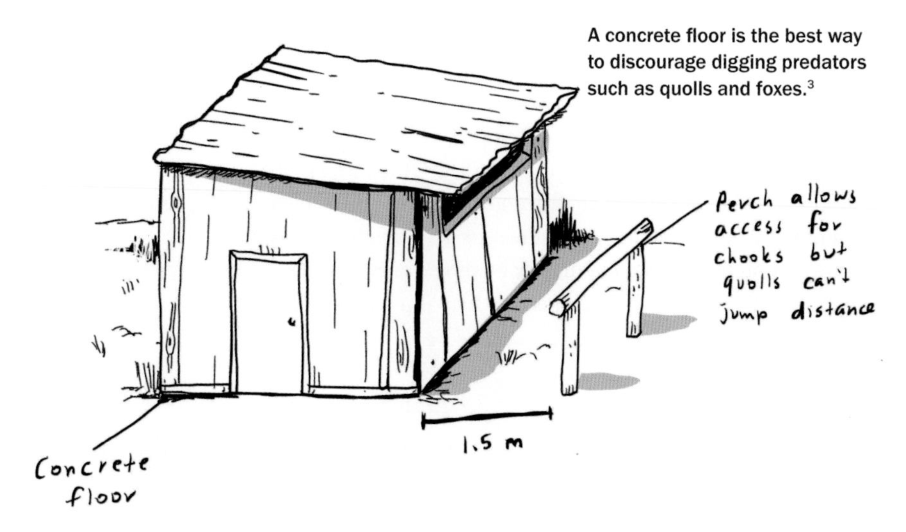

A concrete floor is the best way to discourage digging predators such as quolls and foxes.[3]

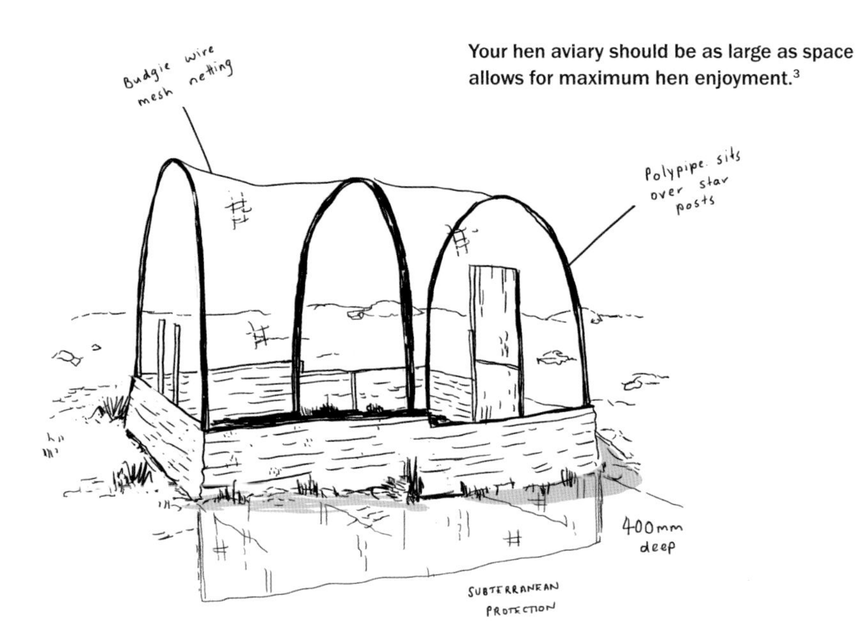

Your hen aviary should be as large as space allows for maximum hen enjoyment.[3]

KIARA MUCCI

Mobile 'chook tractors'

Some gardeners use chook tractors. Without additional protection, these mobile mini-pens leave hens very vulnerable to attacks from quolls or foxes. Peg a protective wire skirt around the base, or only let the hens use the tractor during the day and lock them up in a permanent and secure coop at night.

Free range hens

Keep free-range hens safe by providing protected perches. Isolated trees with a tin collar around the trunk ~1.5 m high will keep hens safe from climbing predators, such as foxes, devils and quolls, but not carpet pythons!

If you're retro-fitting an existing hen house or are using recycled materials to build your hen house, make sure you follow the tips about gaps and the need to extend below ground and secure the materials well so they can't be forced apart. Heavy rain and or stormy weather can cause steel mesh and other materials to move out of place, split or weaken.

What to do if a predator has attacked your chooks

If an attack has occurred, it is likely the quoll or other predator will return now that they have found an easy meal. Any survivors must be taken to the vet for pain relief and an assessment, and further treatment as required.

The remaining hens will be traumatised so the best thing you can do for them is bring them inside, for example in a secure shed or laundry until changes are made to ensure your chicken coop and run are safe.

If you are unsure what was responsible for the attack, seek advice from permaculture or gardening groups in your area, or local wildlife rescue organisation. Try setting up a motion-activated wildlife camera to identify your visitor.

References

1. Department of Environment, Land, Water and Planning (2016) 'National Recovery Plan for the Spotted-tailed Quoll *Dasyurus maculatus*'. Australian Government, Canberra.
2. Glen AS, Dickman CR (2006) Diet of the spotted-tailed quoll (*Dasyurus maculatus*) in eastern Australia: effects of season, sex and size. *Journal of Zoology* **269**(2), 241–248. doi:10.1111/j.1469-7998.2006.00046.x
3. Illustration from: *Protect your chooks and save our quolls*. Saving our Species, Office of Environment and Heritage, Wollongong.

A koala strides across a suburban lawn, intent on their next feed tree.

A visiting koala

KOALAS SPEND MOST OF THEIR lives in trees. But to move from tree to tree, and travel around their home range, a koala must descend to the ground. A few simple measures can ensure our backyards are both safe havens and provide safe passage for a koala on the move.

LOCATION

Koalas occur from Queensland to New South Wales, Victoria and South Australia. Northern koalas have short, silver-grey fur while southern koalas have long, fluffy brown-grey fur.

SEASON

Koalas are most likely to be observed in backyards at the onset of breeding. In the south this is typically from November to March and in the northern areas, such as South East Queensland, from August to October.[1]

SPECIES

Koalas are regarded as one species: *Phascolarctus cinereus*. Genetic studies reveal that there are five distinct sub-populations: Southern Australia, south coast New South Wales, mid-north coast New South Wales, South East Queensland, and the rest of Queensland.[2]

JANINE DUFFY, KOALA CLANCY FOUNDATION

Behaviour

Koalas are renowned for their eucalyptus diets and have been recorded feeding on over 120 species of *Eucalyptus*, *Corymbia* and *Angophora*,[3] and are more likely to be seen in these trees. The eucalypts that koalas prefer vary from region to region, seasonally and even within a koala's home range.

An urban or semi-rural koala has a very specific mental map of suitable feed trees across our gardens, roadsides and reserves, and our backyards may even form part of that koala's home range (or the entire home range of some individuals!). This is especially the case in older suburbs and static rural areas. Radio-tracking of koalas in Redland Shire Council, Qld, as part of their Koala Safe Neighbourhood Program revealed that some koalas spend all their time in people's gardens, like koala ambassador Blake.[4] This means that often when a koala shows up in your garden, it is likely that they know exactly where they are and are not lost. The exception to this is when koalas are caught in a zone of rapid urban expansion and recent habitat destruction has displaced them.

These koalas at Australia Zoo illustrate the remarkable differences in colour and morphology between southern and northern forms. A senior koala on the left from the south is large in size with dark fur, while the younger, northern counterpart is much smaller in stature and has much lighter fur.

AUSTRALIA ZOO

Koalas may also be observed sitting in trees in your garden that are not gum trees. Many koalas like resting in she-oaks (*Allocasuarina*). When it is very hot, koalas will seek shade in large wattles such as blackwoods, cherry ballarts and tea trees, or deciduous trees such as oaks.

Koalas are most active between dusk and dawn, but they will also move during the day to change trees. Movement from tree to tree increases if koalas are stressed by factors such as dogs, road construction or habitat destruction.[5]

Risks

To koalas: The koala exemplifies the challenges facing many species of Australian wildlife that live on the densely populated fertile agricultural lands and low-lying coastal forests on the east coast of the continent – and how threats can combine and then amplify each other. Throughout the koala's range, staff at universities, councils, wildlife hospitals, wildlife rehabilitators and passionate members of community groups hold regular forums to share strategies and solutions to help tackle the multitude of threats facing this much-loved species (see 'Koalas in trouble').

In terms of your garden, a visiting koala is at risk from dog attack, as well as accidental drowning in backyard swimming pools.

To people/property: The claws on a koala's hands and feet are long and curved, allowing them to grip tree trunks with ease, even those with smooth bark. These claws will only be dangerous to humans if you try to handle the koala, and as such koalas are best left to experienced rescuers.

Actions and solutions

First of all, if you have dogs, bring them inside or at least keep them contained on the verandah or in another location well away from the koala. Barking, even by a small dog, can cause stress to a koala. See the wildlife-friendly pet section (pp. 148–156) for more on dogs and wildlife-friendly dog ownership.

While observing your visitor, try to keep a respectful distance from the tree they are in (10 m is a good rule of thumb). If there are a few people observing the koala, try to avoid surrounding the tree.

The next step is to work out whether the koala needs help, which can be tricky! If they are able to, koalas usually climb a tree if they have been hurt, making it hard to know if they have been injured. Calling a wildlife rescue organisation for advice is always recommended, as they can take you through what to look for. Another reason to call someone locally is that koala populations are facing different threats wherever they occur. For example, in South East Queensland disease is the most common reason for koalas being hospitalised,[1] whereas in Victoria it is car strike.[6] If it is a young koala on their own that is the size of a pineapple, this little joey needs care immediately!

It is best to call your local koala rescue group or wildlife rescue organisation straight away if a koala has any of the following symptoms or behaviours, as they may need veterinary attention:

- weeping eyes/conjunctivitis
- 'wet bottom' (their hindquarters are damp and discoloured) = chlamydia or kidney disease
- very thin
- sitting at base of tree during the day.

If you have any suspicion at all that your or a neighbour's dog may have attacked a visiting koala, call a wildlife rescue organisation for help straight away. It can be also very hard to tell if a koala is injured as wounds from a dog attack do not always bleed. Tufts of grey or white fur at the base of the tree can be a clue.

If a koala has a stained brown mark on their chest, this is a scent gland and perfectly normal. Males have scent glands to mark trees in their home range and may come to ground to rub their chest on a tree trunk to do so.

If your koala is okay, you can simply enjoy your visitor until their next visit! Generally, if a koala can make its way into your garden, it can make its way out again.

This pretty-as-a-picture female koala made her way into a house being built on Raymond Island, Vic. As she was healthy, she was left alone and the next day, she moved on.

JANINE DUFFY, KOALA CLANCY FOUNDATION

A koala-safe garden

Universities and local governments in South East Queensland have been working together to deliver innovative community education initiatives using a social marketing approach. A survey of nearly 3,000 residents showed, unsurprisingly, that most people surveyed already care about the koala. Community education thus works best when it promotes the actions that people are already taking and increases understanding of the actions people are required to take in their region. Social media campaigns based around the TV program *The Bachelor*, and plenty of very quick 'how to' graphics have helped people grasp important conservation messages in a fun and practical way.[7]

Dogs

Most fatal dog attacks occur at night, in people's backyards, by dogs that are greater than 10 kg and where the owner has more than one dog.[8] If this describes your situation, the easy solution to this is denning your beloved pet(s). Rather than letting them sleep outside freely in the garden, at dusk secure the dogs in a smaller area such as a garage, laundry or enclosed part of the verandah. The dogs will quickly get used to this routine and love you for it! A small dog who sleeps inside is a wonderfully wildlife-friendly option.

Another way to create a koala-friendly haven at home is to train your dog. The 'Leave It' training program in Queensland works with dog owners to train dogs in wildlife avoidance – helpful for koalas, but also in keeping dogs safe from venomous snakes and kangaroos.[9]

Fencing

Koala-friendly fencing allows a koala to move in and out of your garden with ease. If you have tall, enclosed metal fencing or a paling fence, you can retrofit your fence in less time than it takes to mow the lawn. Simply lean a large wooden plank, pole or very large branch against the fence. Ideally, work with your neighbour to ensure they place another similar climbing option on the other side.

Koala-friendly fencing design includes:

- thick planks that allow the koala to walk across the top of the fence
- small gaps between panels to allow the koala to grip and climb
- trees or sturdy shrubs close to the fence can provide a natural ladder
- provision of a pole or plank leaning at no greater than a 60° angle against the fence can allow for escape.

For more on wildlife-friendly fencing, see pp. 157–162.

Koalas in trouble

All koalas need prime eucalypt habitat so they can obtain good-quality food. When these habitats are cleared for agriculture and urban development, this quite understandably causes increased stress levels in koalas. In fact, habitat destruction causes koalas more stress than any other threat – equal to bushfires.[11] Elevated stress levels in koalas may make the marsupials more vulnerable to diseases such as chlamydia, kidney disease and the koala retrovirus. Chlamydial disease weakens the immune system, and causes blindness and reduced fertility in female koalas.[12]

When habitat clearance is caused by urban development, this multiplies the direct threats to koalas, such as dogs, swimming pools and collision with cars. Studies have shown these threats have a distinct peak in breeding season.[1] Koala breeding season varies from north to south, but it is the ideal time for public education campaigns and increased actions to actively reduce dog attack and car strike particularly in areas of intensifying urban development.

Added to this rather grim combination of threats are the effects of climate change. In 2009 the IUCN included the koala as one of the top 10 species worldwide to be at risk from climate change. Landscape scale bushfires such as the 2019 Black Summer fires kill koalas outright and increase the stress in remaining koalas as their favoured feeding areas are lost.[11]

Climate change increases the frequency and severity of extreme weather events such as droughts. Koalas use behavioural adaptations during heat waves such as tree selection (for a cooler microclimate) and changes in their body posture, but do not cope with heat stress and low water availability at all, leading to local extinctions in drier woodland habitats such as the Pilliga forests in New South Wales.[13]

Climate change also has the strange effect of making the koala's already poor-quality food even lower in nutrients! Scientist have measured the nutritional profile of plants under different levels of carbon dioxide (CO_2) and have discovered that rising CO_2 levels mean trees are growing faster with the result being a lower nutrient level in eucalyptus leaves.[14] Finding enough nutrition in eucalyptus leaves is already a challenge for koalas, meaning they will need to find and eat even more leaves to obtain the same level of nutrients. This will create an even greater reliance on prime quality eucalypt habitat, otherwise forcing koalas to travel further to find enough food, with increased exposure to cars and dogs.

All these pressures have led to the listing of the Queensland, New South Wales and Australian Capital Territory populations of the koala as Endangered under the *Environmental Protection and Biodiversity Conservation Act 1999*. The southern koala populations were not listed, however, as their ecology is not as well studied as northern koalas, and more research is needed to assess and address the threats facing koalas in South Australia and Victoria.[15]

Swimming pools

The name koala used to be considered to mean 'no drink', but they lap water from tree trunks as it rains, and from tiny pools in the nooks and crannies of large eucalyptus branches.[10] During dry, hot periods of no rain, koalas will seek water in swimming pools. Unfortunately, pools without safe access and exit points can be deadly to visiting koalas.

If you are considering putting a pool in, select a design with a lagoon-style, sloped entry. This will enable a visiting koala to climb out of the pool easily if they fall into the water.

It is also easy to retrofit an existing pool. Using a plastic container filled with air and with the lid tightly closed, you can float a thick (at least 5 cm) rope in the pool, attaching it and the float to a nearby tree

References

1. Kerlin DH, Grogan LF, McCallum HI (2023) Insights and inferences on koala conservation from records of koalas arriving to care in South East Queensland. *Wildlife Research* **50**, 57–67. doi:10.1071/WR21181
2. Lott MJ, Wright BR, Neaves LE, Frankham GJ, Dennison S, *et al.* (2022) Future-proofing the koala: synergising genomic and environmental data for effective species management. *Molecular Ecology* **31**, 3035–3055. doi:10.1111/mec.16446
3. Moore BD, Foley WJ (2005) Tree use by koalas in a chemically complex landscape. *Nature* **435**, 488–490. doi:10.1038/nature03551
4. (2022) 'Redland's Coast Koala Conservation Plan 2022–2027'. Redland City Council, Redland.
5. Dexter CE, Appleby RG, Scott J, Edgar JP, Jones DN (2018) Individuals matter: predicting koala road crossing behaviour in south-east Queensland. *Australian Mammalogy* **40**(1), 67–75. doi:10.1071/AM16043
6. Camprasse ECM, Klapperstueck M, Cardilini APA (2023) Wildlife emergency response services data provide insights into human and non-human threats to wildlife and the response to those threats. *Diversity* **15**(5), 683 doi:10.3390/d15050683>.
7. Tkaczynski A, Rundle-Thiele S (2023) Koala conservation in South East Queensland: a shared responsibility. *Australasian Journal of Environmental Management* **30**(1), 48–67. doi:10.1080/14486563.2023.2173320
8. Queensland Department of Environment and Heritage Protection (2012) 'Living with wildlife: koalas and dogs'. Queensland Government, Brisbane.
9. Rundle-Thiele S, Pang B, Knox K, David P, Parkinson J, *et al.* (2019) Generating new directions for reducing dog and koala interactions: a social marketing formative research study. *Australasian Journal of Environmental Management* **26**(2), 173–187. doi:10.1080/14486563.2019.1599740
10. Mella VSA, Orr C, Hall L, Velasco S, Madani G (2020) An insight into natural koala drinking behaviour. *Ethology* **126**, 858–863. doi:10.1111/eth.13032
11. Narayan E (2019) Physiological stress levels in wild koala sub-populations facing anthropogenic induced environmental trauma and disease. *Scientific Reports* **9**, 6031. doi:10.1038/s41598-019-42448-8
12. Quigley BL, Timms P (2020) Helping koalas battle disease – recent advances in *Chlamydia* and koala retrovirus (KoRV) disease understanding and treatment in koalas. *FEMS Microbiology Reviews* **44**(5), 583–605. doi:10.1093/femsre/fuaa024

13. Lunney D, Predavec M, Sonawane I, Kavanagh R, Barrott-Brown G, *et al.* (2017) The remaining koalas (*Phascolarctos cinereus*) of the Pilliga forests, north-west New South Wales: refugial persistence or a population on the road to extinction? *Pacific Conservation Biology* **23**, 277–294. doi:10.1071/PC17008

14. Lawler IR, Foley WJ, Woodrow IE, Cork SJ (1996) The effects of elevated CO_2 atmospheres on the nutritional quality of *Eucalyptus* foliage and its interaction with soil nutrient and light availability. *Oecologia* **109**, 59–68. doi:10.1007/s004420050058

15. Whisson DA, Ashman KR (2020) When an iconic native animal is overabundant: the koala in southern Australia. *Conservation Science and Practice* **2**, e188. doi:10.1111/csp2.188

The brown snake has hooded eyebrows that suggest an angry, aggressive manner. But like all snakes, brown snakes prefer to be left alone.

Venomous snakes

WILDLIFE-FRIENDLY GARDENING PRINCIPLES, SUCH AS providing water for frogs and insects, using logs and stones for lizard lounges, and dense native plantings for small birds and mammals, are all also snake-friendly. While much of our diverse snake fauna is non-venomous, there are a few species that can be dangerous. Thankfully living with snakes is safe and achievable.

LOCATION

Australia-wide.

SEASON

Snakes are most active in the warmer months. Your area may also experience a spike in snake activity and numbers in wet years.

SPECIES

Brown snakes, tiger snakes, red-bellied blacks and copperheads in the southern states. Further north, the diversity and number of species increase and include dozens of non-venomous snakes, such as pythons and tree snakes.

PHOTO: SIMON WATHAROW

Behaviour

Australia is famous for its venomous snakes, most of which are in the Elapid family. This family includes brown snakes (the eastern brown snake, western brown snake and the dugite), tiger snakes (eastern, western and island tiger snakes), copperheads (lowland, highland and pygmy copperhead), the common death adder and around 60 more species. It is a diverse family and colour forms vary considerably even within species. For example, many tiger snakes are a uniform sandy colour without any of the stripes for which the species was named.

Given some snakes are hard to identify, it's usually safest to assume all snakes are venomous. This entry provides advice mainly on venomous snakes, for information on pythons, see pp. 57–61.

Like all of the other animals featured in this book, snakes need a combination of food, water and shelter to survive, which varies from species to species. Brown snakes actively search grass tussocks, compost heaps and trails through the undergrowth looking for prey, most often mice. On the other hand, tiger snakes stay concealed in long grass, looking for frogs. Skinks are eaten by many species of snake, and copperheads and brown snakes will even eat blue-tongue lizards!

So, unfortunately, having blue-tongues in your garden does not keep snakes away, nor do particular species of snake keep other snakes away.[1] Snakes all rely on different prey species and different habitat requirements, so the species in your

This tiger snake has a distinctive striped pattern, but some do not. Snake identification is best left to experts.

SIMON WATHAROW

garden are closely linked to the habitat elements present – a fact that you can use to your advantage in maintaining a garden that is safe for both you and any visiting or resident snake.

Risks

To snakes: Dog and cat attack, lawn mowers, rodenticides. Being killed by humans, car strike.

To people/property: Globally, around 81,000 to 138,000 people die each year because of snake bites, and around three times as many amputations and other permanent disabilities are caused annually. Agricultural workers and children are most affected.[2] Here in Australia with good healthcare and access to antivenoms, the risk really is minimal. The Australian Snakebite Project collated data on snake bites and death over a 10-year period (2005–15), during which a total of 23 people died, with three of these people being snake handlers.[3] In contrast, 19 people died in just 1 year from allergic reactions to bee and wasp stings.[4]

Deaths by snake bite are usually caused by brown snakes and occur most often when people try to kill the snake. While there is a small risk of accidentally treading on a snake, or reaching into their shelter while weeding, the best way to stay safe around snakes is to simply avoid trying to harm them.

There is a risk to cats and dogs. If your cat is an inside cat, then this risk is mitigated, but dogs are a different matter. The tips in this entry should help reduce the risks of a snake attack, but 'Leave It' wildlife aversion training, which was designed to stop dogs

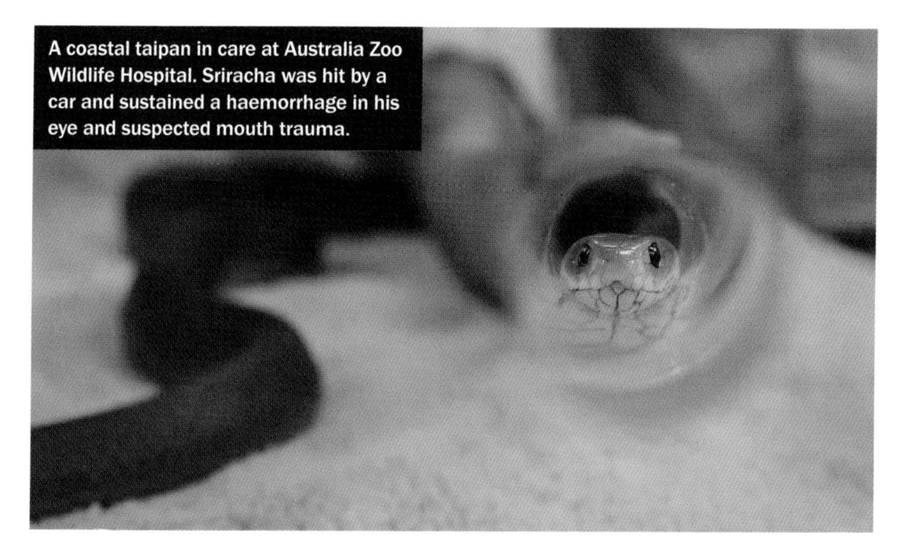

A coastal taipan in care at Australia Zoo Wildlife Hospital. Sriracha was hit by a car and sustained a haemorrhage in his eye and suspected mouth trauma.

attacking koalas, is also ideal for keeping your dog safe from snakes (see pp. 148–156). Pythons can prey on cats and small dogs, which is addressed on pp. 57–61.

Actions and solutions

If you do spot a snake in your garden, remember that often they are just passing through. Simply gather your kids and pets safely inside for a few hours and there is a good chance that this will be a one-off sighting.

If you are regularly coming across snakes in your garden, there are many things you can do to reduce the chances of snake encounters. A simple precaution when sharing your backyard with snakes is to wear long trousers and closed shoes, and use gardening gloves.

Reduce attractants

If you have a rodent problem, try to tackle it! Use a closed compost system, reduce spilled feed if you have chickens and remove any continuous shrubby undergrowth that rodents might like to hide in. This is the most important action you can take to reduce the likelihood of brown snakes using your property, as the introduced house mouse is a major component of their diet.

Snakes are attracted to water, such as pet water bowls and garden ponds, especially on hot days. Place your pet's water inside.

Rock walls, pool edges and concrete slabs (including driveways) are perfect snake habitat. They provide shelter for the animals when it is cool, basking habitat for when it is time to warm up and even provide ideal spots to lay their eggs. Rock walls should be infilled with mortar, leaving no gaps. Other landscaping ideas and actions you can take to reduce potential snake conflict may be found in Table 5.

Table 5. Tips for living with venomous snakes.

Tip	Further information
Design for visibility	Avoid patches of continuous, low-growing groundcovers, long grasses and shrubs. Prune a gap of 30 cm under bushes, or grow shrubs that are devoid of leaves underneath, like many *Eremophila* shrubs. Gravel paths between garden beds are helpful.
Carry out snake-friendly mowing	When mowing or slashing a property, don't start at the boundary or even the middle and then mow towards the house. You will push lizards, snakes and frogs towards the house as they attempt to flee. Instead, finish your boundary first, and then work from the house towards your fence, thus pushing wildlife in the opposite direction away from your house.
If you have a garden pond	Make sure your garden pond is not in the same area you have your pets unsupervised during the day. Keep at least some of the area around the pond paved or with gravel or well mown grass.

Create a safe zone

Create a snake proof area for dogs or kids, such as patio or pergola that uses snake proof wire mesh around the boundary. Another idea is to do the converse – if you have a larger block, create a snake zone away from where you spend your time. Your snake zone can be a patch at the bottom of the garden with water, logs, long grass and corrugated iron.

Snake removal from your garden should be a last resort

Many people's first instinct is to call a snake catcher to remove the snake. While this is a better option than the old days, when the saying 'the only good snake is a dead snake' was standard practice, just like translocated brushtail possums (see pp. 32–37) snakes that are captured and removed from their home range do not fare well at all.

Only 38 (8%) of 464 marked snakes were encountered again after relocation in the city of Darwin, and six were found dead or required euthanasia due to human causes, such as car strike and dog attacks. The researchers concluded that short-distance translocation, to the back of the property, is a better option, particularly with non-venomous snakes.[5]

An Australia-wide survey of the reptile relocation industry found that although most operators agreed that this kind of short-distance relocation was preferable, in practice many were releasing snakes into nearby bushland areas – with multiple snake catchers using the same patch![6] A radio-tracking study of relocated snakes released into a large metropolitan reserve in Melbourne (Westerfolds Park) discovered that this practice may be compounding the problem. Relocated snakes had larger home ranges than resident snakes, and many were leaving the reserve to enter nearby residential properties as they searched for food and shelter.[7]

The reptile relocation industry is largely unregulated, which has disadvantages for both snake catchers and the welfare of the snakes. Contact your local wildlife rescue organisation for recommendations of reputable, welfare-friendly snake catchers in your area.

In Darwin, snake catchers have been elevated to consultants, with snake-catching services jointly managed by the Parks and Wildlife Commission and private contractors. These snake consultants are employed via a competitive tender process, paid for by the government and thus can visit homes and workplaces for free and at all hours.[5] This is an excellent model and would be of great benefit if it was replicated Australia-wide.

What if the snake is in the house or shed?

Immediately call your local snake rescuer or snake catcher for advice. Try to keep an eye on your visitor at all times, as once snakes are out of sight they can be extremely difficult to spot again! Also see pp. 57–61 on dealing with a python in the house.

References

1. Watharow S (2013) *Living with Snakes and other Reptiles*. CSIRO Publishing, Melbourne.
2. Williams DJ, Faiz MA, Abela-Ridder B, Ainsworth S, Bulfone TC, *et al.* (2019) Strategy for a globally coordinated response to a priority neglected tropical disease: snakebite envenoming. *PLoS Neglected Tropical Diseases* **13**(2), e0007059. doi:10.1371/journal.pntd.0007059
3. Johnston CI, Ryan NM, Page CB, Buckley NA, Brown SGA, *et al.* (2017) The Australian Snakebite Project, 2005–2015 (ASP-20). *The Medical Journal of Australia* **207**(3), 119–125. doi:10.5694/mja17.00094
4. Pointer S (2021) 'Venomous bites and stings, 2017–18'. Injury research and statistics series no. 134. Cat. no. INJCAT 215. Australian Institute of Health and Welfare, Canberra.
5. Cornelis J, Parkin T, Bateman P (2021) Killing them softly: a review on snake translocation and an Australian case study. *The Herpetological Journal* **31**, 118–131. doi:10.33256/31.3.118131
6. Derez CM, Fuller RA (2023) The reptile relocation industry in Australia: perspectives from operators. *Diversity* **15**, 343. doi:10.3390/d15030343
7. Butler H, Malone B, Clemann N (2005) The effects of translocation on the spatial ecology of tiger snakes (*Notechis scutatus*) in a suburban landscape. *Wildlife Research* **32**, 165–171. doi.org/10.1071/WR04020

Adult kookaburras are surprisingly fluffy. We can see this young bird is a fledgling by their very short dark bill.

Baby bird out of the nest

EVERY SPRING AND SUMMER, RESCUE organisations and carers are inundated with young birds – but many baby birds brought into care are not in need of rescue! Young birds, known as fledglings, may even walk boldly up to people as if appealing for help. It's important to know when a baby bird actually needs our help.

LOCATION

Australia-wide.

SEASON

Spring and early summer, but in high rainfall years birds may breed all year round.

SPECIES

The most frequent baby bird rescues are for species found in abundance, including Australian magpies, parrots, crows and ravens, kookaburras, tawny frogmouths and honeyeaters such as red wattlebirds.

PHOTO: SASKIA GRANGER

Behaviour

There are different kinds of baby birds, and learning the differences between them can help you determine the best course of action.

When birds such as honeyeaters, parrots and magpies hatch, they are naked, blind and helpless. Whether the nest is a rosella's hollow or a magpie's open cup nest, these chicks stay in the nest and are completely dependent on the parent's care. These are known as altricial chicks. While chicks are in the nest, they are also known as nestlings.

Altricial chicks start off with white fluffy down and then feathers emerge in a tightly rolled sheath (called pin feathers). This first 'feather coat' often also has a few wispy down feathers here and there. The first feather coat of chicks may be a different colour to the adults; for example crimson rosellas are mostly green at this stage, rather than the adult's bright red and blue. By this time, their eyes are open and they are gaining strength in their legs and wings.

When altricial chicks are fully feathered and can see well, two things can happen. Some young birds such as rosellas and kookaburras can fly directly from the nest after some encouragement from the parents, whereas birds such as magpies, currawongs, robins, magpie larks and honeyeaters are able to scramble and hop around, but still

To rescue or to leave alone?

'Baby bird-napping' even occurs with one of our largest birds! In Victoria, November is the time of year when young wedge-tailed eagles fledge. Like smaller birds, young wedge-tails do all of their growing in the nest, so they are the same size as the adults once they leave it. For the first week or two out of the nest, the young eagles are learning to fly, developing their flight muscles and may be observed on the ground. Concerned people may mistake the young eagle for an injured adult and crowd the youngster, attempting to catch it to take it in to a shelter. This may scare the parents away from attending to their chick. An adult eagle can see an animal as small as a rabbit from over a kilometre away, so you can be sure their parents will be watching!

Birds of prey such as hawks, falcon and eagles are best left to experienced rescuers. In the case of the young eagle, a call for advice to a local wildlife rescue organisation would make clear the young bird's behaviour as the wildlife rescue volunteers are able to describe the differences in beak and feather colour between young and adult eagles.

can't fly. The chicks leave the nest – at this stage they are known as fledglings. By the time they leave the nest, fledglings are about the same size as the adult birds. This can cause some confusion, and is further compounded by the fact that a young bird's natural behaviour can be mistaken for an injured adult bird (see 'To rescue or to leave alone?'). Fledglings spend a lot of time on the ground or in low bushes, flapping their wings, doing practice hops and short flights to develop their flight muscles.

Like domestic chickens, when young brush-turkeys, masked lapwings (plovers) and ducks hatch they are already covered in downy feathers. These precocial chicks can see well and even run and feed themselves. Precocial chicks grow their adult feathers in the same way altricial chicks do, but aren't called fledglings as they leave the nest straight after hatching.

Precocial chicks cannot fly for some time after hatching and are very vulnerable to predators, so they are totally reliant on the 'street smarts' of the parent birds. The exceptions are the chicks of the Australian brush-turkey and the orange-footed scrubfowl. In these species, the young emerge from their nest mound and are immediately independent from any parental care!

Risks

To birds: Accurately assessing whether young birds need rescuing is tricky! A young bird that has been taken into care unnecessarily will miss out on the social benefits of being with their parents and siblings, as well as a natural diet. And yet a baby bird that is injured or orphaned needs warmth, a veterinary assessment and specialist care. It is always best to report orphaned birds to your local wildlife rescue organisation and follow their instructions.

To people/property: Raising large numbers of young birds each spring and summer that do not actually require rescue places a huge strain on wildlife rehabilitators and hospitals.

Actions and solutions

If you find an altricial chick

First, observe the chick from a distance and carry out a health and safety check.

Your course of action depends upon the age of the chick. If the chick still has its eyes shut, has no feathers or is naked with down feathers and can't grip onto your finger or a stick, it is still at the helpless stage where it must be in the protection of the nest. If you can locate the nest in a nearby shrub and reach it, you can put the nestling back into the nest. The nestling will be accepted by the parents even after being touched by a human.

If you cannot find the nest, the nest is too high up to reach or the nestling seems injured, call a wildlife rescue organisation for help. You can wrap the nestling in a pillowcase or other cloth to keep it warm. Some sources recommend a hot water bottle filled with warm (not boiling) tap water, but even this may be too hot for the tiny creature, and is best avoided.

If the chick has fully open eyes, downy feathers and can hop around a little bit, but seems very young or unsteady, it is somewhere between a nestling and a fledgling. If you can't find the nest, you can make a makeshift nest for the young bird and hang it in a tree or tall shrub. A 'baby bucket' or 'bird bucket' is a great alternative for birds that live in open nests, such as magpies, crows and ravens, honeyeaters, magpie larks, butcherbirds and tawny frogmouths.

If the chick has fully open eyes, some downy feathers but plenty of normal or adult feathers, usually with a short tail, they are a fledgling. This is the stage at which the young leave the nest under the parents' guidance *before* they can fly. Fledglings spend a lot of time on the ground waiting for parents to come and feed them, and will escape from predators by clambering into trees and shrubs. They often have a

Fledglings are the same size as the adult birds. Marshall the black currawong was likely just out of the nest when he was hit by a car. Here he is pictured in care at Bonorong Wildlife Hospital, Tas.

PETRA HARRIS – BONORONG WILDLIFE SANCTUARY

repetitive begging call, which can range from a regular piping sound in honeyeaters to the whining call of a young magpie or the very odd-sounding wheeze of cockatoos and corellas.

You can help these young birds simply by leaving them alone, and watching them for a while to make sure the parents return. At the fledgling stage, parents may leave their young for hours, so even though it feels counter-intuitive it is preferable to assume that a healthy fledgling is okay and its parents will come back. If necessary, you can make conditions safer for them by guiding them off the road, or herding them into a shrub or bush.

If you find a precocial chick

First, observe the chick from a distance and carry out a health and safety check.

You can help a precocial chick such as a plover by calmly guiding it into the cover of a nearby dense bush or shrub. Retreat a distance so the parents feel it is safe to return to the chick, and ideally they will do so! If the parents don't return after a couple of hours, call a wildlife rescue organisation for advice.

Ducklings are always with their parents, so a lone duckling is in trouble. Avoid the temptation of trying to catch a solo duckling straight away. I tried that once with a chick that had been separated from its mother and siblings via a rock spillway at our local lake. The duckling enacted an incredible predator avoidance speed response, which involved the tiny bird shooting across the water at speed, 10 m or more away from its mother and siblings. I had made the situation worse! A lone duckling will peep loudly and the mother will be able to find it, if she is nearby and there are no barriers between them.

Pacific black ducks will nest in a quiet safe spot in a backyard away from water and then, after the young hatch, lead their brood to water. This can be a treacherous journey, crossing multiple backyards and even roads! Australian wood ducks nest in tree hollows in living and dead standing trees, which means that any tiny fluffy ducklings you observe may have just made their first and final leap out of the nest hollow. As in other duck species, the parents then lead the young to water and safe feeding areas. The ducklings are incredibly vulnerable at this time.

A baby bird needs help in the following circumstances:

- If the baby bird is injured or ill (e.g. falling over on their side, wings drooping, any bleeding, broken wing), or is weak and unresponsive. Sometimes the only sign a bird needs help is the fact they haven't moved in several hours.
- If the fledgling is a 'runner'. Runners are young parrots that have few or no flight and tail feathers. This feather loss is caused by a viral disease known as beak and feather disease. See pp. 64–68 on 'old' cockatoos.

- If the bird has been carried or brought by a cat or dog. Injuries from cats are very hard to see, so the bird will need to be examined by a vet.
- If the parents are dead nearby, the bird needs to go to a wildlife rescue organisation or carer.
- If the baby bird is a nestling (no feathers, bald spots, blind) and you cannot locate the nest.

A common ringtail possum carrying nesting material, Sunshine Coast, Qld.

Possums eating garden plants

RINGTAIL POSSUMS AND THEIR LARGER relatives the brushtails eat mostly leaves and flowers, especially from eucalypts, bottlebrushes and other Myrtaceae species. But possums readily switch to softer fare, including roses and fresh greens from the vegetable garden.

LOCATION

Australia-wide.

SEASON

All year round, but grazing pressure on garden plants may worsen in winter in southern states, as the deciduous trees that possums have come to rely on as a source of food go dormant.

SPECIES

Common ringtail possum, western ringtail possum, common brushtail possum, mountain brushtail (bobuck) and short-eared brushtail possum.

PHOTO: ETHAN MANN

Behaviour

As discussed previously, there are higher densities of brushtail possums in urban areas than in surrounding bushland (see pp. 32–37). It is a similar story with the smaller ringtail possums, as our backyard habitats may offer more food, shelter and water resources than the bushland reserves nearby. Both common ringtails and western ringtails do rather well in well-watered gardens with a dense shrub layer, ideal for both foraging and nesting. Possums that may occur in backyards vary from state to state – and in some areas there can be more than one resident species. For example, my bush block in central Victoria has mountain brushtails and common ringtails regularly, with the occasional common brushtail possum passing through. Refer to Table 6 for possible backyard possums in your state.

Like so much Australian wildlife, brushtail possums use tree hollows for shelter, but can also take to roof spaces or nestboxes. Ringtail possums (both western and common) are unusual in that they build a spherical nest out of twigs, finely shredded bark and leaves called a drey. Dreys are usually tucked into tangled vines, low trees or tall shrubs, and look like an untidy bird's nest, but wholly spherical. Sometimes dreys are built within roof spaces, or under solar panels.

The western ringtail is similar in appearance to the common ringtail of eastern Australia, but sadly the western ringtail is critically endangered. They once occurred across the south-west region of Western Australia, from Perth to Albany. Today, the possums occur at just 10% of their former range, due to land clearing for agriculture and urban development.[2] Researchers thought that western ringtails living in gardens must be using nearby bushland at least some of the time, so some researchers decided to test this assumption by radio-tracking 18 ringtails living in backyards in Albany, WA.

Several months later, the researchers were surprised to discover the radio-tracked ringtails remained exclusively in people's gardens – irrespective of how close they were to a patch of bushland. What's more, the ringtails were observed using 'exotic'

Table 6. Possums from state to state: mammal distribution matrix from *The Field Companion to the Mammals of Australia*.[1]

Species	Qld	NSW	Vic	Tas	SA	WA	NT
Common ringtail possum							
Western ringtail possum							
Common brushtail possum							
Short-eared brushtail possum							
Mountain brushtail possum							

plants more frequently than indigenous plants for foraging, movement around their range and for their day-time refuges.[3] Whether this is from a dearth of native plants and some canny switching on the part of the possums, or from the fact the possums actually prefer exotics, we don't know. What this does mean is that all residential gardens in the western ringtail's range are potentially critical habitat for this possum – so if you are in Western Australia, we recommend rolling out the red carpet for your little ringtails and letting them eat all the roses they desire!

One would think we know so much about the marsupials in our backyards. But we are learning new aspects of possum ecology all the time, thanks in part to camera trapping studies. It has long been known that brushtail possums are partial to bird nestlings and eggs, and it was thought that ringtails are wholly vegetarian. Camera trapping at a nest revealed that common ringtails also prey on eggs – in that case from an eastern yellow robin nest.[4] The ringtail possum is described as strictly arboreal, but new research reveals that ringtails, like koalas, regularly come to ground to move from tree to tree, and camera trap images show the ringtails foraging on the ground.[5]

Risks

To possums: The decline of western ringtails provides some insight into what happens to possum populations under pressure and is a warning that although 'common' ringtails are not threatened now they are subject to a similar suite of threats.

Ringtails are very vulnerable to attack by pets – cats and dogs. During the 6 months of the above study, two of the 18 possums were killed by pets, and one killed by vehicle strike – another very common cause of death for these possums.[3] The ringtails' willingness to use new and human-made structures has created a terrible problem for ringtails who build their dreys in tin roofs or under solar panels. During heat waves, the drey heats up unbearably, and when the possum flees to seek shade, they burn their paws on the roof, which may reach temperatures of 59°C.[6]

Possums are also prone to electrocution, as they use power lines to avoid facing the dangers on the ground – but this can be fatal if they reach for another line while on one.

To people/property: Flower, fruit and vegetable damage.

Actions and solutions

Make sure it actually is a possum!

Black rats are superlative climbers and often possums are blamed for rat damage. One way to tell is to look for droppings or scats. Brushtail possum scats are broadly oval in shape, often clumped together, and vary in colour depending upon the most

Wildlife photographer Ethan Mann used a camera trap to photograph this common ringtail possum navigating a garden fence on the Sunshine Coast – also capturing an image of himself inside the house opening the fridge!

ETHAN MANN

recent plant foods consumed. Ringtail possum scats are small cylindrical pellets ~1 cm long, and uniform in shape and colour as they are composed mainly of finely chewed leaves. Rat droppings are ~1 cm long, and often pointed at the ends. If it turns out you have a rodent problem, see pp. 49–56.

Wallabies such as the black wallaby are also partial to roses and other flowers. They stand up quite tall and use their forepaws to pull down surprisingly large branches to access the flowers, fruit and foliage.

Erecting a wildlife camera to capture images of your forager may help with identification.

Try landscaping/planting techniques

Possums have well-used movement paths to get around the garden safely each night. Often these paths are backyard fences. Try placing a path adjacent to the fence line and planting on the other side of the path to the fence, as this means roses and fruit trees are much less accessible.

Some gardeners say that possums do not like aromatic plants such as chrysanthemums, mint bushes, geraniums, rosemary and lavender, so try planting mini-hedges around your rose bushes. Spiky plants such as spiny grevilleas or hakeas may also form a protective barrier – and may deter wallabies, too.[7]

For canopy trees such as fruit trees, you could try reducing possum access to your trees by trimming branches so that trees and shrubs are at least 1.5 m apart.

These measures are unlikely to completely stop hungry and determined possums that have taken a liking to particular plants, so another method that may seem counterintuitive is to plant more food for your possums!

Find out what possums in your area like to eat. Planting local, native and evergreen species provides possums with food during lean times such as winter when deciduous trees are dormant, and may reduce browsing on new buds in spring. In Sydney, for example, ringtail possums love spotted gums (*Corymbia maculata*), heath banksias (*Banksia ericifolia*), and silky tea-trees (*Leptospermum myrsinoides*). In Western Australia, western ringtails eat peppermint leaves (*Agonis flexuosa*), other native leaves such as jarrah (*Eucalyptus marginata*), marri (*Corymbia calophylla*) and local native plants such as *Melaleucas* and wattle species (*Acacia* spp.). Each state will have a different suite of local species, and any plant suitable for ringtails will be suitable for brushtails also.

Do possum repellents work?

Home remedies and commercially available products aim to work by repelling possums by taste, or by smell. Taste-based repellents include strong, smoky black tea, garlic spray, fish sauce, tabasco sauce and quassia chips – and the commercial product D-ter®. Testing revealed that hungry brushtail possums were just as likely to eat apple pieces with these repellents applied.[8]

Smell-based repellents include Charlie Carp® fertiliser, blood and bone and commercial products D-ter®, Keep Off® and Scat®. Olfactory repellents have had a little more success than taste-based ones, but all must be reapplied after rain, and success seems to vary widely depending on different possum populations and their preferences. Blood and bone, and stockings filled with dog hair, are other commonly used remedies, but these have not been tested.

Some possum deterrents are simply toxic and should be avoided, such as smearing Vics® VapoRub on plastic cling film around the base of trees or hanging naphthalene balls in your fruit trees.

Separating your prize plants via barriers

On existing trees

Tie mesh bags around ripening fruit – as a bonus, this will keep off hungry parrots too.

Try possum-proofing your trees using a tree collar made by affixing a sheet of metal or polycarbonate plastic around the tree trunk – just make sure the tree collar is of sufficient girth to let the tree grow each year.

Barriers on dedicated flower, fruit and vegetable beds

The floppy fence: Possums are excellent climbers, and ringtails can even use their white-tipped tails as a fifth limb. The floppy fence deters possums via a floppy top that sways and makes it very hard for the possum to climb.

The polycarbonate slippery fence, also sheet metal: Another climbing barrier, a sheet of material such as polycarbonate plastic or sheet metal can prevent the possums from purchase on the smooth surface.

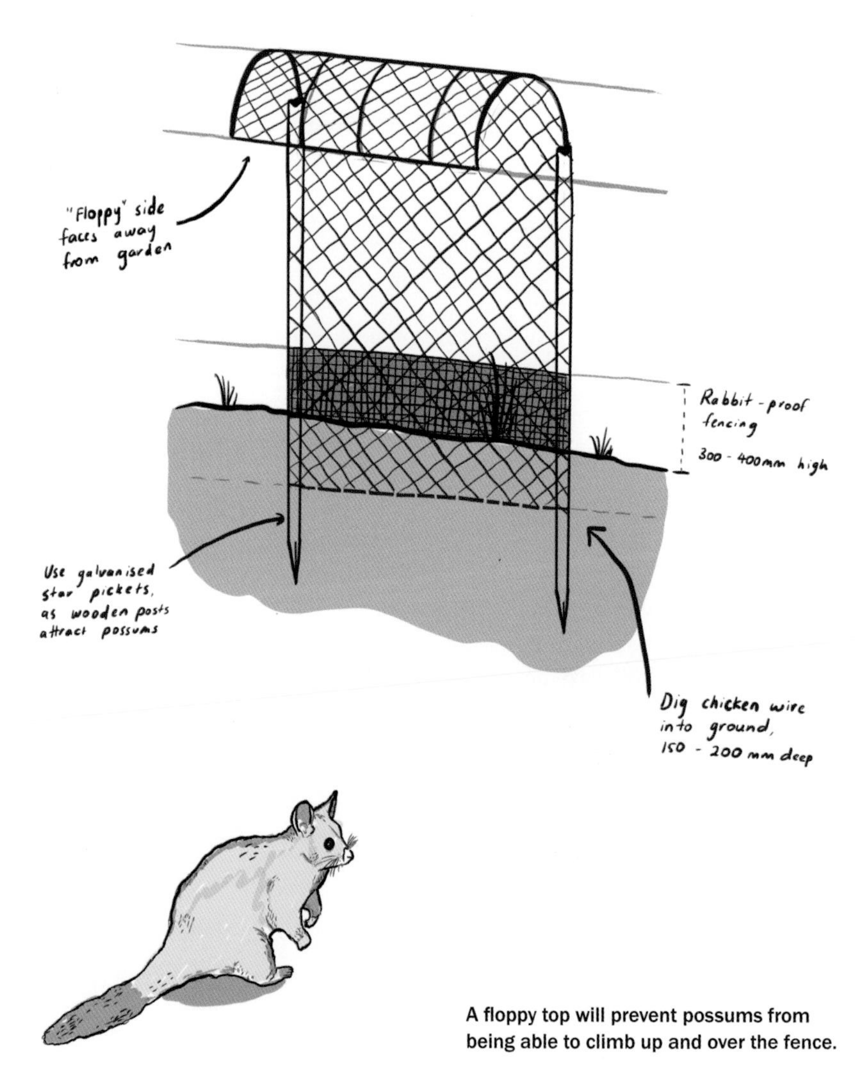

A floppy top will prevent possums from being able to climb up and over the fence.

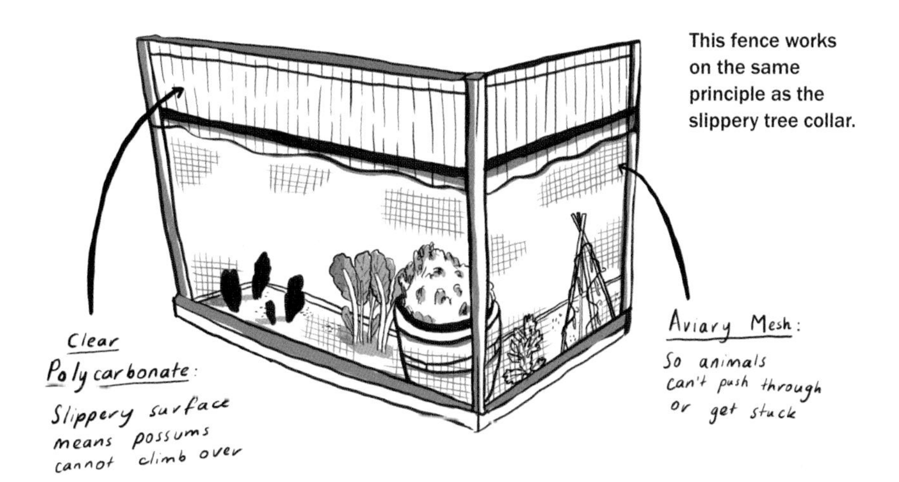
This fence works on the same principle as the slippery tree collar.
Clear Polycarbonate:
Slippery surface means possums cannot climb over
Aviary Mesh:
So animals can't push through or get stuck

KIARA MUCCI

The total enclosure! Total enclosure of your vegetable gardens and orchard is the biggest investment in time and resources – but protects your plants from *all* of your hungry backyard visitors at once, including possums, wallabies, deer and parrots. Contact your local permaculture or gardening group for design ideas. I have seen some incredible enclosures made with star pickets, polypipe and wire netting.

References

1. Van Dyck S, Gynther I, Baker A (Eds) (2013) *The Field Companion to the Mammals of Australia*. New Holland Publishers, Sydney.
2. National Environmental Science Program Threatened Species Research Hub (2019) 'Threatened Species Strategy Year 3 Scorecard – Western Ringtail Possum'. Australian Government, Canberra. <http://www.environment.gov.au/biodiversity/threatened/species/20- mammals-by-2020/western-ringtail-possum>.
3. Van Helden BE, Close PG, Stewart BA, Speldewinde PC, Comer SJ (2021) Critically Endangered marsupial calls residential gardens home. *Animal Conservation* **24**(3), 445–456. doi:10.1111/acv.12649
4. Yu N (2023) Predation of eastern yellow robin nestlings from the same nest by an eastern ringtail possum and a pied currawong. *Australian Field Ornithology* **40**, 166–169. doi:10.20938/afo40166169
5. Van Helden BE, Close PG, Stewart BA, Speldewinde PC, Comer SJ (2020) Going to ground: Implications of ground use for the conservation of an arboreal marsupial. *Australian Mammalogy* **42**(1), 106–109. doi:10.1071/AM18053
6. Siah DS, Busselton Veterinary Surgery. Presentation at 2023 Australian Wildlife Rehabilitation Conference, Perth Western Australia.
7. Horsfall M (2014) *Australian Garden Rescue*. CSIRO Publishing, Melbourne.
8. Temby I (2005) *Wild Neighbours: The Humane Approach to Living with Wildlife*. Citrus Press, Sydney.

Latrice the southern cassowary poses on the verandah, while the photographer's parents keep their distance!

Cassowary visits

THE SOUTHERN CASSOWARY IS AN occasional visitor to gardens in Far North Queensland. These formidable fruitarians know when particular fruits are ripe across their home range, which could include your tropical backyard! There are many things you can do to ensure you have a cassowary-friendly garden and feel safe when they drop by.

LOCATION

Australia's southern cassowary population is split in two, with the southernmost grouping in the Wet Tropics, which is well-studied, and a lesser known population on Cape York Peninsula.[1]

SEASON

Cassowary visitation depends on the species of plants in your garden. Some plants fruit all year round, while others are seasonal. If your garden has a water source, then you may get visitors year round.

SPECIES

The southern cassowary *Casuarius casuarius* is also found in Papua New Guinea, but the cassowary we see in Australia is a distinct subspecies known as *C. casuarius johnsonii* – the southern or double-wattled cassowary.[2]

PHOTO: PATRICK TOMKINS

Behaviour

Southern cassowaries' preferred habitat is dense tropical rainforest, especially lowland rainforest. They are very hard to see in the rainforest gloom, despite their large size, brightly coloured wattles and head piece known as a casque. Cassowaries are most often observed on the edges of the rainforest, such as roadsides. They may also be seen in cane fields and eucalypt forest next to rainforest areas. And of course, parks and backyards.

Rainforest plants of the Wet Tropics rely on a range of animals to pollinate flowers, as well as eat fruits and disperse seed. Flying-foxes and tube-nosed fruit bats play a role in seed dispersal, as do forest birds such as pigeons and catbirds, but only the cassowary is large enough to consume iconic rainforest species such as blue quandong (*Elaeocarpus* genus).[3]

Unlike the fruit that we modern humans eat with smaller 'pips' and plenty of fruit flesh, much rainforest fruit is less fleshy: for example, the blue quandong only has a thin layer of blue flesh around a large seed. The cassowary thus eats a lot of fruit each day and their droppings or dung work as seed distribution agents.

Many plant species rely on passing through the cassowaries' gut to germinate. A study of cassowary diet found that cassowaries consumed fruits of 238 species![3] Local rainforest nurseries even invite locals to bring in a sample of cassowary dung from their properties and then propagate the plants that grow from these.

A cassowary known as Queen Ru feeds on quandong by the creek.

PATRICK TOMKINS

As they travel widely across their home ranges, the birds are also able to disperse the seeds over large distances (more than 2 km),[4] helping to keep the rainforest healthy by intermingling plant genes across the landscape.

Cassowaries stand up to 2 m tall, with males weighing up to a hefty 55 kg and the females are even heavier at up to 76 kg.[2] These large birds need to eat a *lot* of fruit to obtain the energy they need.

Risks

To cassowaries: The natural predators of cassowaries include crocodiles, pythons, dingoes and quolls, but many of these animals prefer to prey on their eggs and chicks rather than tackling an adult cassowary. Unfortunately, these dangers are minimal compared with human-made threats faced by cassowaries today: destruction and isolation of patches of rainforest habitat, car strike, dog attack and predation by and competition with feral pigs. In the Wet Tropics many habitat restoration projects, rainforest buyback programs and traffic calming initiatives are focused on reducing threats to cassowaries. Compounding these threats is the predicted increase in the frequency and intensity of tropical cyclones due to climate change.[4] Cassowaries are long-lived birds with a slow breeding cycle, so every adult cassowary counts at this stage!

If you are lucky enough to have your backyard as part of a cassowary's home range, having an enclosed verandah and a run for your dog is recommended, so that your pet can enjoy a view of the garden without entering the area where the cassowary may visit. A cassowary is likely to visit when a particular tree is fruiting (see Table 7) and you may need to keep your dog restrained for the amount of time that the cassowary is visiting – so why not have a permanent safe area for your pet?

As with koalas (see pp. 103–110), choosing a smaller breed of dog (under 10 kg) is cassowary-friendly, but in the Wet Tropics this size dog will need to be protected from pythons!

Table 7. The cassowary calendar.[5]

January–March	April–June	July–September	October–December
Adult males are seen with large young while females are solitary. Birds roam extensively because food is scarce and eat almost anything they can find, including dried droppings.	This is still a difficult time for the birds. Adults begin courting in May/June and the young are evicted from the home range. This is a hazardous time for immature birds learning to fend for themselves in competition with intolerant breeding adults. Many juveniles die during this period.	Fruiting trees in the lowlands are producing reliable food – many favoured species are in full fruit at this time. Eggs hatch during the maximum fruiting period. Males are very protective of their chicks and humans should give them a wide berth.	Males are moving around with their striped or brown young. There is no shortage of food. This is a very important period when birds build up reserves to cope with inevitable shortages in the new year.

A tourist hand-feeds a southern cassowary at Etty Bay, Far North Qld, despite signs nearby stating there is a $5,000 fine for doing so.

ETHAN MANN

Feeding cassowaries directly is *not* a good idea and in fact is subject to a very large fine from the Queensland Government. Exotic fruits such as bananas do not have the correct nutrients that cassowaries need compared to their rainforest food, but what is much worse is that cassowaries become habituated to humans and seek out human habitation – thus increasing their exposure to threats such as car strike and dog attacks. Feeding cassowaries also substantially increases the risk of attacks to humans. Of 150 attacks against humans in 1985–99, 75% were by cassowaries that appeared to be expecting or soliciting food.[6]

To people/property: Although it is said that cassowaries are naturally curious, when a cassowary has abundant natural habitat with plenty of fruiting trees over their home range, they are generally wary of humans and will tend to keep their distance. Unprovoked attacks are extremely rare.

Even a bird as large and formidable as the cassowary gets tricked by window reflections and may attack their reflection, mistaking it for an intruder in their territory. In the above-mentioned study, cassowaries kicked windows or doors in eight incidents, six of these resulting in broken glass or screens.[6]

Actions and solutions

If you are standing on a verandah or in a well-frequented area of your garden and a cassowary arrives, it is likely that the cassowary knows that you are there and all will be fine if you do not approach.

Table 8. Cassowary behaviours from low-intensity to increasingly agitated.[7]

Behaviour	Description	How to respond
Pecking	Ritualised pecking at ground followed by a fixating stare at intruders.	Once the bird starts staring at you this is the best time to back away slowly. Try to put a tree between you and the cassowary.
Ritualised preen	Bird faces intruder at an angle of 75° or stands at lateral view 90°. Preens chest area stopping to stare pointedly at intruders, and may flare their vestigial wings.[8]	As above
Strut	Bird walks slowly, lifting legs higher than is necessary, with neck extended upward.	As above
Frantic movement	The strut can turn frantic, turning to the right and left while walking in short straight lines past the intruder. The bird will always turn towards the intruder and then move past the intruder again. Cassowary may be cornered.	As above
Charging, running, chasing, pushing	These aggressive behaviours most often happen in birds that are used to being fed.	Cassowaries can run faster than humans so government advice is to stay still. Protect your front with a bag or backpack.
Kicking, jump or jump-rake	Although kicking with powerful legs and clawed inner toe is what cassowaries are renowned for, this is very rare behaviour	Last resort behaviour, and ideally no reader will have to deal with this!

If times are tough for your local cassowaries, it is a good idea to give your visitor a wider berth as there may be a small chance cassowaries will defend a particularly coveted food source, such as a fruiting tree in the garden. Five per cent of the attacks against humans were by cassowaries that appeared to be defending their food or feeding areas, including mangoes, custard apples figs and tomatoes.[6] Periods of increased aggression can occur following a cyclone, in drought years or if habitat destruction for development is reducing the availability of rainforest fruit locally.

Fathers with chicks of any age also need to be given plenty of space, as 6% of attacks against humans were by males defending chicks.[6]

One researcher studied the signs of stress or aggression in cassowaries, so that people could read the signs and back off before an attack could happen.[8] The descriptions of cassowary behaviour in Table 8 may be useful for people with cassowaries in their garden or for overly avid photographers!

References

1. Westcott DA, Metcalfe S, Jones D, Bradford M, McKeown A, *et al.* (2014) 'Estimation of the population size and distribution of the southern cassowary, *Casuarius casuarius*, in the Wet Tropics Region of Australia'. Report to the National Environmental Research Program. Reef and Rainforest Research Centre Limited, Cairns.

2. Latch P (2007) 'National recovery plan for the southern cassowary (*Casuarius casuarius johnsonii*)'. Report to Department of the Environment, Water, Heritage and the Arts, Canberra. Environmental Protection Agency, Cairns.
3. Westcott DA, Bentrupperbäumer J, Bradford MG, McKeown A (2005) Incorporating patterns of disperser behaviour into models of seed dispersal and its effects on estimated dispersal curves. *Oecologia* **146**, 57–67. doi:10.1007/s00442-005-0178-1
4. National Environmental Science Program Threatened Species Research Hub (2019) 'Threatened Species Strategy Year 3 Scorecard – Southern Cassowary'. Australian Government, Canberra. <http://www.environment.gov.au/biodiversity/threatened/species/20-birds-by2020/southern-cassowary>.
5. Wet Tropics Management Authority (2012) 'Tropical topics: Cassowaries', <www.wettropics.gov.au/site/user-assets/docs/Cassowaries.pdf>.
6. Kofron C (1999) Attacks to humans and domestic animals by the southern cassowary (*Casuarius casuarius johnsonii*) in Queensland, Australia. *Journal of Zoology, London* **249**, 375–381. doi:10.1111/j.1469-7998.1999.tb01206.x
7. Keller M (2009) Behaviour of Southern Cassowary (*Casuarius casuarius*) towards intruders. *The Sunbird* **39**(2), 49–53.
8. Patrick Tomkins, personal communication (6 October 2023).

An echidna digs for termites and ants by a water tank, Etty Bay, Queensland.

Echidna in the backyard

INSTANTLY RECOGNISABLE, THE ECHIDNA TRUNDLES through a wide variety of habitats across the country – from woodlands to forests to arid areas. So a surprising number of Australians may come across a spiky visitor but, unlike a possum, a brush turkey or other more generally well-accepted urban species, the sight of an echidna in the backyard causes consternation for many echidna lovers.

LOCATION

Australia-wide.

SEASON

All year round.

SPECIES

The echidna's formal name is the short-beaked echidna. Echidnas range in colour from dark brown to a light sandy colour.

PHOTO: ISTOCK.COM/FLURIN SEGLIAS

Endangered echidnas

The Kangaroo Island subspecies *Tachyglossus aculeatus multiaculeatus* is the most well-known echidna population thanks to decades of work by Dr Peggy Rismiller and subsequent generations of researchers.[1] Echidnas are surprisingly difficult to study! Of course, they are easy to see when you come across one, but very hard to locate if you are looking for a particular individual over a large area.

Echidna populations across Australia are regarded as reasonably secure, but Dr Rismillers's studies have shown that the Kangaroo Island echidna population has declined precipitously in recent years. The Kangaroo Island echidna is now classified as endangered – largely due to the combination of car strike, feral cats and the impact of the destruction of half the island's bushland habitat in the 2020–21 Black Summer fires. Young echidnas are particularly vulnerable. Researchers are concerned that if the Kangaroo Island echidna is in trouble, then it is likely that more work needs to be done to make sure our mainland and Tasmanian echidna populations are stable.

Behaviour

What is an echidna doing in your backyard? The answer lies in the fact that the echidna's home range is larger than you might think. An oft-quoted figure is a home range of 50 ha; an area of some 50 rugby fields! My bush block of 2.1 ha represents only a fraction of the possible home range of the echidnas we see here occasionally.

The home range is where an echidna chooses to feed, rest and breed. Adults may remain faithful to this area – one study showed a female was still using the same area after a period of 13 years.[2] But studies of wild echidnas have shown that the size of an echidna's home range can vary widely – between habitats, male and female echidnas, and whether echidnas are young or mature.

Echidnas range across a wide range of habitats, from all natural areas to cities. Participants in the Echidna CSI citizen science program have submitted photos of echidnas while walking in their local bushland, visiting the beach and even at the snow! Project leader Dr Tahlia Perry and her team were also surprised by the large number of echidna records submitted from in and around major metropolitan cities.[3]

The echidna visiting your backyard could be a repeat visitor – perhaps they visited a full 10 years ago? Your backyard could also be visited by several different echidnas, as their home ranges can and do overlap with one another.

Just like koalas, who move through our backyards as they travel around their home ranges, the most important thing we can do is ensure the echidna's safe passage.

Risks

To echidnas: In the backyard the biggest threats to echidnas are pet dogs and cats. In your neighbourhood, echidnas are vulnerable to car strike.

To people/property: None.

Actions and solutions

Rest assured, if a visiting echidna has found a way into your backyard, they will find their way out again! Despite their short legs and unusual boxy shape, echidnas are surprisingly agile and strong, and can swim, climb and burrow with ease. They usually move on within 24 hours.

The first and most important step is to remove dogs and cats from the area, if necessary. As echidnas often travel and feed at night, they are another reason to keep your pets inside during night-time hours.

The only defensive mechanism an echidna has is to dig down into the soil or sand with a circular digging motion, using their long claws and unique backward-facing hind feet. These rapidly dug burrows protect their vulnerable snout and underbelly, while leaving the protective spines exposed.

Resist the temptation to keep checking on your echidna – the longer they are left alone, the sooner they will realise the coast is clear and that they can keep moving.

The echidna will be able to move on without assistance. There are reports of concerned people capturing the echidna and transporting to the nearest bushland area; this is not necessary and will cause undue stress to the echidna.

Possible puggles

Another reason to leave the echidna alone is that it may be a mother with a young echidna or puggle located in a nursery burrow nearby. These nursery burrows can be in a natural space that has been widened such as under a log, or rocks and may have a long, curved entrance way.

Puggles hatch from a small leathery egg, and then remain in the mother's pouch until they are ~50 to 60 days old. By this time, the puggle starts growing fur and spines, and it is time for it to leave the pouch and stay in the nursery burrow. The mother may leave her puggle in the nursery burrow for up to 5 days while she travels over her home range, foraging for up to 18 hours a day and consuming as much food as possible. Upon her return, a mammoth 2-hour feeding session ensues, with the puggle lapping nutritious milk from the milk patch on her belly. The puggle can consume 40% of their body mass in one session![4]

After feeding, the mother leaves the sated puggle and backfills the entrance as she goes. Keeping the entrance closed in this way keeps the young puggle safe, but can also make it nigh on impossible for humans to know if they have a nursery burrow in their garden!

Echidna safe bush backyard tip: if burning brush piles is part of your seasonal fire clean-up, make sure you create and burn the piles in the same day. If you pile up branches and leaf litter and then leave them until next fire season, a mother echidna may think your brush pile is the perfect site for a nursery burrow - with devastating consequences for a young puggle that may be inside.

References

1. Department of the Environment (2023) *Tachyglossus aculeatus multiaculeatus* in Species Profile and Threats Database, Department of the Environment, Canberra. <https://www.environment.gov.au/sprat>.
2. Nicol SC, Vanpé C, Sprent J, Morrow G, Andersen NA (2011) Spatial ecology of a ubiquitous Australian anteater, the short-beaked echidna (*Tachyglossus aculeatus*). *Journal of Mammalogy* **92**(1), 101–110. doi:10.1644/09-MAMM-A-398.1
3. Perry T, Stenhouse A, Wilson I, Grützner F (2022) EchidnaCSI: engaging the public in research and conservation of the short-beaked echidna. *Proceedings of the National Academy of Sciences of the United States of America* **119**(5), e2108826119. doi:10.1073/pnas.2108826119
4. Baker A, Gynther I (2023) *Strahan's Mammals of Australia Fourth Edition*. New Holland Publishers, Sydney.

Part 3: Helping wildlife in trouble

A sign implores drivers to take care on a narrow road through forest in regional New South Wales.

PHOTO: ISTOCK.COM/LYNNE BECLU

Injured, orphaned or unwell animals

Make sure you are safe and call for help

Any wildlife rescue needs to consider the safety of the rescuer first and foremost! If you are on a road, watch out for traffic. When pulling your car over, make sure it is entirely off the road, as you may be liable for any accidents caused if not. If you are in an extreme weather event such as storm, flood or bushfire situation, call for help first rather than risking your life. If the animal is caught on a powerline, call for help.

Remove any sources of stress, risks or threats to the injured animal, such as dogs, children and even well-meaning people wishing to take a photograph.

When you call for help you will be prompted to state your specific location. Your location is vitally important, as the animal (should it survive) will need to be released back within its own territory or home range. This is particularly vital for lizards and turtles, which may have small home ranges, and for animals such as possums, they may even benefit from you knowing which tree they were beneath.

If you can, waiting until the rescuer arrives is very helpful, as it saves valuable time while the volunteer searches for the animal. If you are on a long stretch of road, look for the unique numbers on the power poles – a good way to reference a location. Photographs, GPS readings or screen shots of your map on your phone can also be a good option to help pinpoint the location and can be shared when the animal is taken into care.

Not knowing the exact location of where some species are found can reduce the chance of their survival when released, or even mean they are not able to be released at all – in short, the better the detail, the better the outcome.

Do not approach some wildlife

Consider whether it is safe to approach the injured or unwell animal. Some kinds of Australian wildlife can bite, scratch or kick, and are best handled by an experienced and trained wildlife rescuer. Wildlife Health Australia advises to avoid approaching snakes, monitor lizards (also known as goannas), bats (flying foxes or microbats), adult kangaroos and wallabies, raptors (eagles, hawks, falcons and owls) and marine mammals (whales, dolphins, seals and turtles).[1] Koalas, wombats, cassowaries, brush turkeys, Tasmanian devils and large quolls are also best left to experienced wildlife rescuers.

The above 'do not approach' list is rather long, and only really leaves possums, bandicoots, magpies and other birds, frogs and small- to medium-sized lizards to be approached. Although safer, these animals can still cause injury, or be injured themselves, if handled without the appropriate equipment or experience.

Call for help and advice straight away

Even if you have had experience with looking after domestic animals or humans, first aid and care for Australian wildlife are very different and require specialist care. Sick and injured wildlife mask their pain and injuries at all costs in order to appear normal, so don't wait to see if the animal improves or the animal's situation changes. Sick and injured animals are more likely to be attacked by predators, or even members of their own species, so if the animal seems at all out of sorts or unwell, it is likely they are in serious need of assistance.

Remember: there is no harm in calling for help even if you are unsure as to whether the animal needs assistance. All wildlife rescue organisations are happy to respond to a call about an animal who may be in trouble and if the animal turns out to be okay, then that is a great outcome for everyone involved!

Who you should call varies from state to state, and whether it is a weekday, weekend or public holiday, but an internet search should be able to locate your nearest help. By the time of publication, the IFAW wildlife rescue app, which will have listings of local wildlife rescue organisations in your area, should be rolled out across Australia. Veterinary surgeries, wildlife rescue organisations, wildlife hospitals and even government organisations such as the police may be able to help. Save your local rescue numbers in your phone so you always have them handy – or if travelling somewhere new, pre-search for the local wildlife help and save the number in your phone.

Once on the phone, explain any visible injuries that you can see on the animal without touching it.

If the advice you are given is to take the animal to a nearby vet and you are able to, please do so! There is a mistaken belief that if you take injured wildlife to a vet 'it will just be euthanised'. Certainly, the capacity of vets to respond to wildlife patients varies considerably from practice to practice, but if you have been advised the animal needs veterinary attention, it is likely they need an X-ray or specialised treatment that only a vet can provide.

It is a sad fact that long-term studies of the fate of wildlife brought into care at wildlife hospitals in Queensland and New South Wales have shown that over 50% of wildlife are euthanised after examination due to the extent of their illness or injuries.[2,3,4] Compassionate euthanasia is always preferable to an animal left untreated in pain and distress.

Keep the animal safe and quiet

Whatever wildlife rescue organisation you contact will be able to advise you on the best methods of capturing and securing the animal, but a torch, pillowcase, gloves, beanie, towel and a sturdy cardboard box with ventilation holes and a lid are the most useful tools. Byron Bay Wildlife Hospital have a fantastic kit for sale on their website, but you can easily assemble your own to keep in your car or spare room at home. Unfortunately, much wildlife trauma occurs outside of business hours, on weekends and busy holiday periods. The advice summarised in Table 9, while no substitute for guidance from a wildlife rescue organisation, may assist in keeping the animal safe until help arrives.

In some situations, you won't be able to catch the animal, but can still take action while waiting for help (see Table 10).

Once the animal is safe, make sure it is kept in a dark and quiet room, with no pets, children or noise from the television or radio. The animal will be in shock, and a safe space is what they need until specialised help is available. If you are driving an animal to a vet or wildlife hospital the same applies – keep the car quiet and calm, without music or loud talking, and leave your dog at home.

Minimise handling and avoid offering them food or water

Reducing the stress experienced by the animal is your top priority and, although it seems counter-intuitive, avoid patting or stroking the animal as you would a dog or cat. While it seems like a caring thing to do, an animal such as a koala or wallaby is not tame and such physical contact and your close proximity are stressful. They might think you are attacking or preparing to prey on them! Minimum handling is the best approach for stressed wildlife.

Table 9. Types of wildlife and how to secure them if necessary.

Type of animal	How to catch and secure
Bird (excluding birds of prey)	Wearing gloves, use a towel to gently cover the bird, including its head. Lift them into a cardboard box, ensuring their wings are in their normal resting position – be careful of their beak biting through the towel – and cover it with the lid. Once they are in the box, do not transport them wrapped up in the towel. Instead, make sure the towel covers the bottom of the box so they have something to grab onto and don't bounce around in the box while being transported. For baby birds, see pp. 117–122.
Lizards, dragons and turtles (not monitor lizards)	Make sure the reptile is not a snake! Wearing gloves, use a towel to capture the animal. A lot of reptiles can give a nasty bite, so always pick them up with a towel so they bite it before you! If it is a larger dragon, handle it with one hand under the belly and the other holding its tail. Hold turtles with their rear well away from your body as they have a defensive spray mechanism that smells very bad. Place the animal into a cardboard box with air holes and cover with the lid. **Note:** Shingleback lizards are prone to becoming flyblown almost immediately, so will need to be moved into a protected area away from flies straight away.
Frogs	Moisten your hands, and wear moistened gloves if possible before touching any frog as their skin is very delicate and may tear if handled with a dry surface. Place in a plastic container with some leaves, damp grass and cover with a lid with air holes.
Echidna	Wearing good-quality leather gloves, wrap a thick (or folded over) towel around the echidna to pick it up. Place the animal in a tall plastic container, such as a bin (an injured echidna can escape from a cardboard box), with a secure lid with air holes for ventilation. Layers of towels should be used to cover the bottom of the plastic container. In hot weather, one of the towels should be dampened with cool water to keep the temperature below 25 °C.
Large possums	If found at home, large possums on the ground can be covered with an upside-down laundry basket, with a towel or sheet laid over it secured with a weight. If on the road, wear gloves and use a towel to capture the possum, making sure to cover its head.
Small possums	Small possums may be captured by gently placing a towel over the animal, including its head, and then placing it in a cardboard box covered with a lid.
Bandicoots (quenda)	Cover the animal with a towel and use both hands to pick it up. These animals are particularly subject to stress, avoid grabbing its tail, as the skin may easily be stripped from the tail. Place the animal in a securely tied bag or a cardboard box.
Possums and gliders (orphaned)	If the young animal is fully furred, remove it from its mother using a towel and wrap it in pillowcase (consider taking the mother's body with the baby on its back, as this is where it will be most comfortable and will reduce the chance of the baby running or biting). If not furred, leave the animal on its mother's teat and place the dead mother and orphan in a box together. Do not try to take them off the teat as their delicate mouthparts can be damaged.
Kangaroos and wallabies (orphaned)	If the young is fully furred, remove it from its mother and wrap it in a pillowcase. If not furred, leave the animal on its mother's teat and call for help.

Table 10. Fence hangers and koalas.

Type of animal and situation	What to do
Flying fox, owl, glider caught on barbed wire fence	Avoid touching the animal. Cover the animal with a towel or erect some shade with a tarp or umbrella. Call for help.
Injured or unwell koala	Avoid touching the animal. If they are on the ground, place a box or basket over the koala to prevent them from moving. Call for help.

The exception to this is when caring for a wallaby, kangaroo or wombat joey. If a joey is still in its mother's pouch, it will not be able to create its own body heat. Place the joey in a pillowcase turned inside out (to prevent them suckling and ingesting threads and labels), then put a beanie around the pillowcase and place it down your jumper to keep the joey warm.

Stressed, injured animals do not need food – imagine if you had just been in a car accident and instead of administering first aid the paramedics tried to feed you a delicious meal. Like stroking a rescued animal, wanting to feed the animal is well intentioned but quite the opposite to what the animal needs, which is a dark and quiet space that minimises the shock and trauma while waiting for professional help.

The wrong kind of food can also injure an animal. For example, a rainbow lorikeet's tongue is a delicate brush-tipped organ for lapping nectar and pollen from flowers. Being fed bird seed may damage their tongue.

Surprisingly, water can also pose a danger to injured or stressed animals. During bushfires and extreme heat events, the media is filled with images of people pouring water down the throats of thirsty koalas. But in natural conditions, koalas drink with their head bent down, using a lapping motion with the tongue, like a cat. Water poured from a water bottle can end up in the koala's lungs, resulting in a disease called aspiration pneumonia, which may be fatal. In extreme heat or bushfire situations water may be required, but your wildlife rescue service can advise.

Do not keep the animal and care for it yourself

Some of us may recall our grandparents, or uncles and aunties, raising orphan magpies and wallabies, or keeping a galah that had suffered a broken wing, and remember the joy and connection that resulted from these experiences. However, wildlife rescue is a specialised role that requires training, support from other carers and a wildlife licence. For some of these animals, the care given by untrained people works out okay, but for others it doesn't.

Wildlife care organisations are frustrated by well-meaning people trying to care for an animal for a few days or even months and then bringing the animal into care, when by this time it is often severely malnourished. In most cases the animal needs to be euthanised.

Today it is accepted that Australian wildlife such as marsupials, parrots and lizards require specialist diets, socialisation with their own kind and expert care to thrive. Taking the animal to a vet, wildlife carer or wildlife hospital is always the best option for the animal, and in fact is a legal requirement under the state wildlife laws.

References

1. Wildlife Health Australia (2021) 'What to do if ... : fact sheet August 2021'. Canberra, <https://wildlifehealthaustralia.com.au/Portals/0/Incidents/What_to_do_if_Sick_injured_dead_wildlife.pdf>.
2. Heathcote G, Hobday AJ, Spaulding M, Gard M, Irons G (2019) Citizen reporting of wildlife interactions can improve impact-reduction programs and support wildlife carers. *Wildlife Research* **46**(5), 415–428. doi:10.1071/WR18127
3. Taylor-Brown A, Booth R, Gillett A, Mealy E, Ogbourne SM, *et al.* (2019) The impact of human activities on Australian wildlife. *PLoS One* **14**(1), 28. doi:10.1371/journal.pone.0206958
4. Kwok ABC, Haering R, Travers SK, Stathis P (2021) Trends in wildlife rehabilitation rescues and animal fate across a six-year period in New South Wales, Australia. *PLoS One* **16**(9), e0257209. doi:10.1371/journal.pone.0257209

This little bichon frise poses little threat to koalas, but still requires supervision to ensure he does not harm bandicoots and other smaller animals.

PHOTO: ISTOCK.COM/PETER VAHLERSVIK

Wildlife-friendly pet ownership

We love our companion animals. Nearly half of Australian households own at least one dog, and a third own at least one cat. I am equally passionate about wildlife and dogs, and have experienced the conflicted feelings that occur when a beloved pet manages to chase or even kill wildlife. Many dogs and cats retain strong hunting instincts and when allowed to fulfil these urges may have devastating effects on wildlife in backyards, parks and nearby bushland. The good news is that a very pampered pooch or pussy cat is also the most wildlife-friendly pet.

Wildlife-friendly dog ownership

Feral cats and pet cats and their combined effect on Australian wildlife has received a lot of media attention of late and rightly so – but less widely discussed is the potential for our pet dogs to cause harm. Pet dogs have a disproportionately higher impact on some kinds of wildlife, including koalas,[1] wallabies, blue-tongues, pademelons, brushtail possums, echidnas,[2] bandicoots, blue-tongues,[3] shinglebacks, snakes and even cassowaries.[4] Dogs tend to attack around the head and neck, and sometimes these attacks include vigorous shaking of the prey. The strength of the dog's jaws and the muscularity of the attack in general mean that often animals' injuries are so extensive that they require euthanasia.

Chasing also harms wildlife. When a dog barks and chases without catching and or physically touching the animal, it may seem like there is no harm done. The dog is simply 'having a good time'! Unfortunately, being chased by dogs can cause marsupials (e.g. ringtail possums and quenda) to drop their pouch young as they flee. Barking and chasing also increases stress in animals that are already suffering (e.g. koalas with chlamydia). Repeated dog chasing also disrupts the daily activities of wildlife such as finding food, raising young and defending territories.[5]

Sometimes, just the presence of dogs can have an impact, even when walked on a lead. Researchers carried out a study comparing the number of birds recorded at 90 sites in woodlands around Sydney: half at sites where dogs were regularly walked, and the other half at sites where dogs were excluded (National Park areas). When dogs were walked researchers recorded a 25% reduction in bird diversity and a 35% reduction in abundance of birds compared to sites when just humans walked on these trails. Moreover, the study recorded that the most popular dog-walking trails had on average 10 dog walkers and their dogs per hour![6]

The study was carried out in bushland on the edges of Sydney with a suite of naturally occurring woodland bird species. Do dogs have the same effect on tougher urban adapted birds in city parklands – and by extension our gardens? Magpie-larks are often found in cities and initial research shows these birds are able to tell when a dog is on a lead, and react accordingly. If the small dog used in the study was on a lead, the birds mainly just walked away, and if the dog was off lead the magpie-larks flew away – a much more costly response energetically.[7] These studies show us that it is likely that our pet dogs have an effect on our backyard birds, and thankfully the intelligence and resourcefulness of these birds mean that we can take steps to further reduce our impact.

Many of the entries in this book have featured tips and advice for coexistence between our beloved dogs and wildlife, and will be summarised here.

Choose wildlife-friendly breeds and mixed breeds

Ideally, adopt or choose a toy or companion breed instead of hunting dog. Even small hunting breeds such as terriers may have strong hunting instincts and your local lizards, bandicoots and other small creatures will need to be protected from them. A dog under 10 kg in weight is less likely to attack and kill animals such as koalas and wallabies.[1]

Keep your dog inside at night with you

A large proportion of our wildlife is nocturnal – if your dog or dogs are inside with you at night then your resident wildlife such as possums, and those passing through such as koalas may use your garden as habitat undisturbed. This practice is increasing. Dogs are now more likely to be kept indoors only – from 13% in 2019 to 17% in 2022 – though the majority (71%) continue to be kept both indoors and outdoors.[8]

Another idea is to have an enclosed patio or denning area that is separate to any trees or habitat for your dogs to sleep in each night.

Use fencing to contain your dog

You can also use wildlife-friendly fencing to create pet and wildlife zones in your garden. A well-fenced dog run or enclosed area will keep wildlife such as koalas and bandicoots out of the dog zone, and will also protect your dog from snakes. A wildlife exclusion fence can be constructed out of sheet metal or via a thick wooden paling fence, with no gaps (see the section on wildlife-friendly fencing on pp. 157–162).

Try training your dog

I once had a lovely experience staying at a holiday house in Wye River, Vic. The property owner had trained his two cattle dogs to stay calm and not react to koalas,

This encounter is likely to be stressful for both dog and koala!

WILDLIFE VICTORIA

so myself, my nephews, the owner and the two dogs sat quietly together and watched a large healthy koala climb down a tree, stroll casually across the length of the garden and saunter up another tree, completely unbothered by his audience!

In South East Queensland dog attack is an ongoing threat to the endangered koala population, and so the social marketing team at Griffith University have created a free online training program called 'Leave It'.[9] This training not only helps wildlife like koalas – it could also save your dog's life, as you can call the dog away from threats such as kangaroos (see pp. 89–95) and venomous snakes (see pp. 111–116).

Use a lead – even at home!

Using a lead comes in handy even while you are at home. Our bush block is unfenced, so we use a long retractable lead to take our small schnoodle from the house to the car on warm days when the copperhead is active, when there are ducklings in the garden and when a kangaroo or two is visiting.

Wildlife-friendly cat ownership

Pet cats (*Felis silvestris catus*) are descended from the Near Eastern wildcat (*Felis silvestris lybica*).[10] Unlike the differences between a Great Dane or a schnauzer (*Canis lupus familiaris*) and a grey wolf (*Canis lupus*), the physical and genetic differences between house cats and their cat ancestors are minimal.[11] Cats are amazingly adaptable – with the one species equally at home on the couch indoors dining on 'Fancy Feast' or roaming the desert subsisting on rabbits and reptiles. For this entry we focus on owned or pet cats, where owners provide for their cats, may take measures such as desexing or microchipping, and may have some control over their movements.

Just like dogs, some kinds of wildlife are more affected by cat predation. Animals under 4 kg are the rule – with ringtail possums, feathertail gliders, microbats and small birds such as fairy-wrens particularly at risk, as well as reptiles including young blue-tongue lizards. Often when an animal is attacked by a cat, the wounds are very hard to see, as a cat's incisors are so sharp they can penetrate the skin and underlying organs without visible damage. If there is any suspicion of cat attack, the bird or lizard will need examination by a vet or wildlife carer.

Having said the above, taking cat owners on a journey to wildlife-friendly cat ownership is not achieved simply by hating cats and their hunting instinct, demonising cat owners or reciting statistics. Instead, participatory programs that work with pet owners are much more effective.[12,13] These programs help cat owners better understand their cat's behaviour and offer reassurance and support for cat owners wishing to transition their cats to an indoors lifestyle.

The Cat Tracker project was deliberately designed to engage cat owners and assist them to make informed decisions about their cats, via participation in a large scale pet cat tracking project in Adelaide, SA. GPS tracking collar data from 428 pet cats were collected on their movements for a week. Most cats, 75%, had home ranges under 2 ha, while a few cats had much larger home ranges (3% of cats had home ranges over 10 ha), leaving a median home-range size of just over 1 ha.

The majority of participants found these results surprising and interesting, and all the project participants stated that their cats went further than expected.[12]

Participants were surprised and concerned about the number of times their cats were crossing the road on their adventures. The average number of roads crossed per day was 4.8 and the median was 3.4! Cats with home ranges of over a hectare crossed more roads, and were in more fights with other cats.[13]

The Adelaide study tracked mainly in urban/suburban areas. The Tasmanian Cat Tracker project, led by the Cradle Coast Authority, was curious whether cats on larger more rural properties wandered further. The study had a small sample size, with just 20 cats but it included larger, rural properties as well as urban – and found that cats on rural properties roamed even further, although the median size of the home ranges was similar to the Adelaide cats at 1.4 ha. One cat, Ollie, wandered over 7 ha. His home range was three times larger than the home range of any other cat in the project.[14]

Both Cat Tracker programs helped owners get to know their cats, their movements and hunting habits – and resulted in a greater understanding of the benefits of containment, both at night and during the day.

Containment: a safe cat and the most wildlife-friendly option

Like dogs, cats are being increasingly kept indoors only – from 36% in 2019 to 42% in 2023, with a corresponding decrease in those who allow their cats both indoors and outdoors (from 58% to 51%).[8] This is great news for both cats and wildlife!

It is generally accepted that dogs must be kept on an owner's property and be leashed or under control if their owners are out and about. But up until the last 10 years or so one or more cats that come and go as they please from their owner's property has been the norm. This lifestyle allows cats to fulfil their natural exploring, hunting and social behaviour with other cats, and owners feel good that their cats are living their best lives. But freedom to roam may come at a great cost to the cat's welfare!

As the tracking projects have demonstrated, many pet cats cross the road multiple times as they roam and may be killed by cars. Cats that are unrestrained are more frequently injured, especially as a result of fighting with other cats, and may suffer from disease, including exposure to feline enteritis.[15] Some cats prey upon venomous snakes while they are out and about.[3]

Many cat owners are concerned about the potential harm that may befall cats, but at the same time worry that containment is actually cruel to their cats.[16,17] The good news is that as long as the cats have access to sunning spots, toys and games, garden views and perhaps a patch of cat grass these cats will live long and fulfilling lives!

Campaigns such as Zoos Victoria's and RSPCA Victoria's Safe Cat, Safe Wildlife, TassieCat and RSPCA Australia's Safe and Happy Cats website have a wealth of advice and support for both existing cat owners and those about to get a cat.[17]

Some councils have programs which support residents who wish to contain their cats via educational workshops and material subsidies on wildlife-friendly gardening and the building of outdoor cat runs, also known as 'catios'.[18,19]

Total containment eliminates the risk of disease spreading from cats to wildlife. Finn, the young wombat pictured on p. 28, was orphaned when his mother contracted toxoplasmosis, which occurs in cat droppings. Finn was also affected and needed weeks of intensive care at Bonorong Wildlife Hospital, Tas.

How about partial containment – keeping kitty in at night?

Keeping a pet cat contained at night could be considered a good compromise, and a step-wise approach to full containment.[20] This practice reduces some of the risks cats face, including road crossings. Partial containment stops pet cats from preying upon nocturnal wildlife such as ringtail possums, microbats and feathertail gliders.[3] In the case of the critically endangered western ringtail, partial containment could be vital for the future of the species.[21]

The most wildlife-friendly approach is total containment. Keeping your cat in at night may not work as well as you may think as some cats are adept at 'sneaking out' while their owners sleep. Out of 177 cats whom owners believed were inside at night, 69 (39%) were recorded out on night adventures![13]

Containment at night does not protect diurnal or day living wildlife, with small birds such as fairy-wrens and reptiles such as young blue-tongue lizards particularly vulnerable to cat attack. And some pet cats have a habit of attacking venomous snakes.[3]

But my cat only catches a few animals a year

The animals that a cat owner sees are the trophy catches – the small proportion the cat chooses to bring home. Seven studies (non-Australian) compared the predation rates of pet cats as revealed by scat or cat-borne video to the prey return rates; on average, only 15.1% of killed prey was brought home. This means potentially 85% of animals killed by pet cats are not observed by their owners![17]

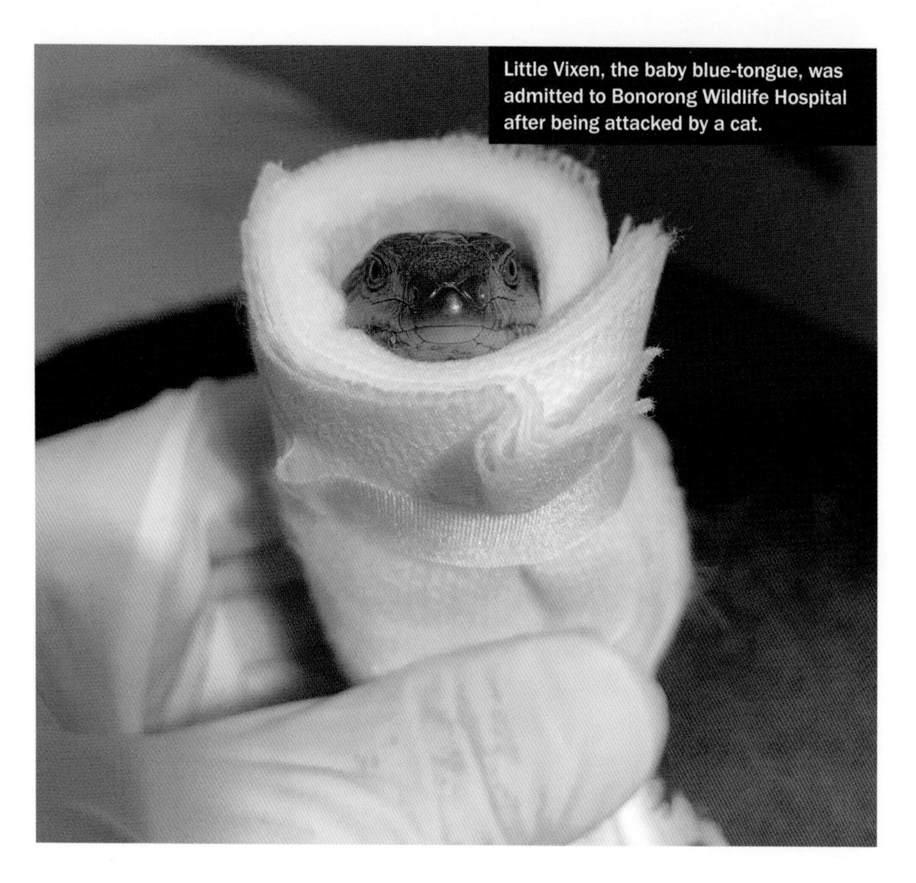

Little Vixen, the baby blue-tongue, was admitted to Bonorong Wildlife Hospital after being attacked by a cat.

MATTHEW CLEMENT – BONORONG WILDLIFE SANCTUARY

What about CatBibs and cat collars known as Birds-be-Safe?

Like partial containment, these measures reduce predation on wildlife but they do not eliminate it.

Alone or together with bells one study reported that CatBibs stopped 81% of cats from catching birds, 45% from catching mammals and just 33% of cats from catching frogs and lizards. These figures are based upon prey that was returned to their owners.[21]

A study of Birds-be-Safe collars, colourful cat collars much like a hair scrunchie, also reported a decrease in hunting birds in 78% of cats in the study. Again, this figure calculated from prey brought home. Reductions in other prey groups were not measured.[22]

Like partial containment, these predator deterrents do not eliminate the risks faced by cats as they wander, and a safe and happy cat inside is the most wildlife-friendly solution. These deterrents could be a compromise solution to suggest to a neighbour, friend or relative who is living in or near high conservation areas and who is reluctant to transition to any kind of containment for their pet cat.

References

1. Queensland Department of Environment and Heritage Protection (2012) *Living with wildlife: koalas and dogs*. Queensland Government, Brisbane. <https:/ environment.des.qld.gov.au/wildlife/animals/ living-with/koalas/threats>.
2. Heathcote G, Hobday AJ, Spaulding M, Gard M, Irons G (2019) Citizen reporting of wildlife interactions can improve impact-reduction programs and support wildlife carers. *Wildlife Research* **46**, 415–428. doi:10.1071/WR18127
3. Taylor-Brown A, Booth R, Gillett A, Mealy E, Ogbourne SM, *et al.* (2019) The impact of human activities on Australian wildlife. *PLoS One* **14**(1), e0206958. doi:10.1371/journal.pone.0206958
4. Latch P (2007) 'National recovery plan for the southern cassowary (*Casuarius casuarius johnsonii*)'. Report to Department of the Environment, Water, Heritage and the Arts,. Environmental Protection Agency, Canberra.
5. Doherty TS, Dickman CR, Glen AS, Newsome TM, Nimmo DG, *et al.* (2017) The global impacts of domestic dogs on threatened vertebrates. *Biological Conservation* **210**, 56–59. doi:10.1016/j.biocon.2017.04.007
6. Banks PB, Bryant JV (2007) Four-legged friend or foe? Dog walking displaces native birds from natural areas. *Biology Letters* **3**, 611–613. doi:10.1098/rsbl.2007.0374
7. Barnett SC, van Dongen WFD, Plotz RD, Weston MA (2023) Leash status of approaching dogs mediates escape modality but not flight-initiation distance in a common urban bird. *Birds* **4**(3), 277–283. doi:10.3390/birds4030023
8. Animal Medicines Australia (2022) 'Pets in Australia: a national survey of pets and people'. Barton.
9. Rundle-Thiele S, Pang B, Knox K, David P, Parkinson J, *et al.* (2019) Generating new directions for reducing dog and koala interactions: a social marketing formative research study. *Australasian Journal of Environmental Management* **26**(2), 173–187. doi:10.1080/14486563.2019.1599740
10. Driscoll CA, Menotti-Raymond M, Roca AL, Hupe K, Johnson WE, *et al.* (2007) The Near Eastern origin of cat domestication. *Science* **317**, 519–523. doi:10.1126/science.1139518
11. Crowley SL, Cecchetti M, McDonald RA (2020) Our wild companions: domestic cats in the Anthropocene. *Trends in Ecology & Evolution* **35**(6), 477–483. doi:10.1016/j.tree.2020.01.008
12. Roetman P, Tindle H, Litchfield C (2018) Management of pet cats: the impact of the cat tracker citizen science project in South Australia. *Animals (Basel)* **8**(11), 190. doi:10.3390/ani8110190
13. Roetman P, Tindle H, Litchfield C, Chiera B, Quinton G, *et al.* (2017) *Cat Tracker South Australia: Understanding Pet Cats Through Citizen Science*. Discovery Circle, University of South Australia, Adelaide South Australia. doi:10.4226/78/5892ce70b245a.
14. Oorebeek M, Pauza M (2021) 'Cradle Coast Cat Tracker report.' Tasmanian Government, Cradle Coast Authority, Burnie.
15. Tan SML, Stellato AC, Niel L (2020) Uncontrolled outdoor access for cats: an assessment of risks and benefits. *Animals (Basel)* **10**(2), 258. doi:10.3390/ani10020258
16. Legge S, Woinarski JCZ, Dickman CR, Murphy BP, Woolley LA, *et al.* (2020) We need to worry about Bella and Charlie: the impacts of pet cats on Australian wildlife. *Wildlife Research* **47**(8), 523–539. doi:10.1071/WR19174
17. Van Eeden L, Hames F, Faulkner R, Geschke A, Squires ZE, *et al.* (2021) Putting the cat before the wildlife: exploring cat owners' beliefs about cat containment as predictors of owner behavior. *Conservation Science and Practice* **3**(10), e502. doi:10.1111/csp2.502

18. Nou T, Legge S, Woinarski J, Dielenberg J, Garrard G (2021) 'The management of cats by local governments of Australia'. Threatened Species Recovery Hub, Brisbane.

19. Linklater WL, Farnworth MJ, van Heezik Y, Stafford KJ, MacDonald EA (2019) Prioritizing cat-owner behaviors for a campaign to reduce wildlife depredation. *Conservation Science and Practice* **1**, e29. doi:10.1111/csp2.29

20. National Environmental Science Program Threatened Species Research Hub (2019) 'Threatened Species Strategy Year 3 scorecard – western ringtail possum'. Australian Government, Canberra. <http://www.environment.gov.au/biodiversity/threatened/species/20-mammals-by-2020/western-ringtail-possum>

21. Calver M, Thomas S, Bradley S, McCutcheon H (2007) Reducing the rate of predation on wildlife by pet cats: the efficacy and practicability of collar-mounted pounce protectors. *Biological Conservation* **137**(3), 341–348. doi:10.1016/j.biocon.2007.02.015

22. Pemberton C, Ruxton GD (2020) Birdsbesafe® collar cover reduces bird predation by domestic cats (*Felis catus*). *Journal of Zoology* **310**(2), 106–109. doi:10.1111/jzo.12739

A grey-headed flying-fox hangs tangled in urban fruit tree netting. Despite being rescued, the netting had cut the circulation to its wing for too long and so the bat had to be euthanised. Prahran, Vic, Australia.

PHOTO: DOUG GIMESY

Entanglement

Entanglement describes the many ways wildlife may come into strife via encounters with fences, fruit tree netting, discarded fishing tackle, bottle cap rings and other rubbish. Delicate bat and bird wings and glider membranes are caught on barbed wire points, entrapping the unfortunate animal. Bats, birds and even snakes struggle when caught in fruit tree netting and become further entangled, suffering from shock, exposure and dehydration. Hooks, fishing line and plastic rings become ensnared and cause injuries particularly around the head and feet.

Entanglement is one of the cruellest of fates and easily prevented by taking a few simple measures.

Fence entanglement

Barbed wire – best removed or avoided

To our eyes, a barbed wire fence appears obvious. But at times of low visibility such as dusk or in misty conditions, wildlife such as owls, bats and gliders simply do not see the barbed wire fence and become entangled. Flying-foxes are particularly prone to being snagged on barbed wire, especially if located near a food or water source – with sometimes hundreds of flying foxes impaled on one fence in one night![1,2] Not only flying-foxes are affected: a study in 1999 collated records of 62 species involved in barbed wire entanglement, including squirrel and sugar gliders, barn owls, southern cassowary, small birds such as eastern spinebill and silvereye, ducks, parrots, herons and mammals such as koalas, microbats, tube-nosed fruit bats, bettongs and the Tasmanian pademelon.[3]

Barbed wire discourages larger stock animals such as cattle from pushing through fences to reach to the grass on the other side. This is a very specific use, yet barbed wire is still considered by many as part of a typical fence and used widely in smaller bush block and even semi-rural suburban settings – where there are no cows!

What to do if you have a property with fencing that is not wildlife friendly

Once fencing is erected, it may remain for decades, and depending on the size of your property it can be expensive to redo. Happily there are several ways to retrofit an existing fence to make it safer for wildlife. See Table 11 for options.

Make sure any new fencing you erect is wildlife-friendly

The most wildlife-friendly fence is a post and rail fence, made of timber. This kind of fence allows wildlife free movement from property to property. Wallabies, wombats, echidnas, bandicoots, blue-tongue lizards and kangaroo joeys are able to pass under the fence. Kangaroos can safely hop over the fence without fear of entanglement, and koalas may use their formidable claws to scale the fence.

Table 11. Wildife-friendly fencing solutions.

Options to consider	Comments
Total removal.	Perhaps you don't need a fence at all! There are other ways to demarcate property boundaries. A hedge of banksia, wattles or lilly pillies looks wonderful and is great wildlife habitat. A line of rocks can delineate a property boundary and provide lizard habitat at the same time. In larger properties, a series of small 'keep out' signs can ensure you have privacy, while allowing wildlife free movement.
Remove the top strand of barbed wire.	95% of all entanglements occur on the top strand. Removing the top strand of the fence effectively removes much of the risk without removing the entire fence.
Make the top strand of the fence more visible.	This can be done by adding high visibility white tape along the length, or even old CDs or DVDs tied on at intervals.
Cover a section of the barbed wire with polypipe.	If barbed wire fence is next to a favoured food source such as a flowering shrub or fruit tree, place split poly pipe over the area of the fence near the shrub, thus encasing the barbed wire in poly piping and rendering it safe.

Post and rail fences are great for delineating property boundaries, but if you want to keep your dogs in your garden, and not running on the road, the next best alternative is a simple wooden post-and-wire fence, with the top strand white for enhanced visibility. This white strand is visible at night and helps animals to safely clear the fence. You can use old electric fence white tape or buy a white roll of electrical wire.

An alternative to barbed wire for larger landholders with stock has been created: 'barbless barbed wire'. It does sound strange – you may ask, why not use just plain wire? But this innovative product has the same tensile qualities as barbed wire and comes in the same format, so landholders and fencing contractors can switch products while working with the same tools and using the same fence building techniques.

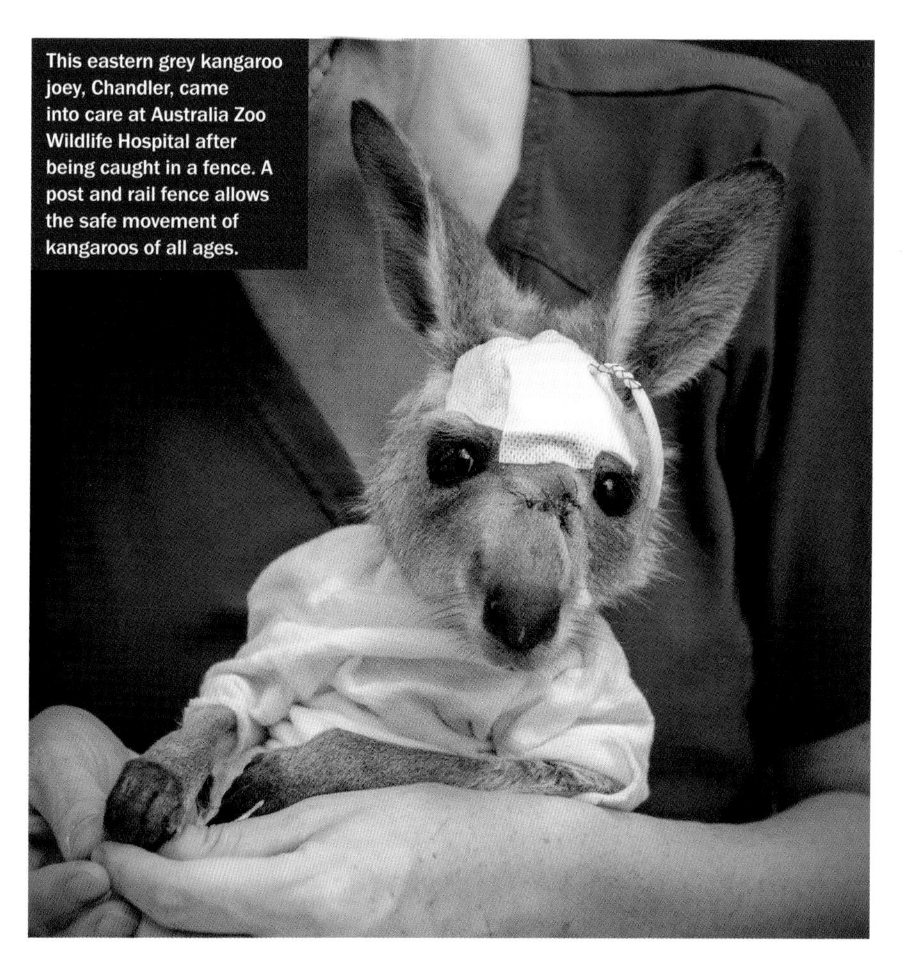

This eastern grey kangaroo joey, Chandler, came into care at Australia Zoo Wildlife Hospital after being caught in a fence. A post and rail fence allows the safe movement of kangaroos of all ages.

Fruit tree netting entanglement

Protecting fruit trees seems like such an innocent endeavour but unsafe fruit tree netting is equal to barbed wire in the dangers it presents to wildlife, especially flying-foxes.[4] The Wildlife Friendly Fencing and Netting website has a running tally of species suffering from unsafe fruit netting entanglements, including rainbow lorikeets and other parrots; reptiles such as brown snakes, pythons, eastern water dragons and lace monitors; eastern snake-necked turtles and frogs; and even the echidna![2]

Happily, there is an easy fix – if you choose to net your fruit trees, make sure you use wildlife-friendly fruit tree netting.

Unsafe fruit tree netting has an open weave with a mesh size greater than 1 cm square. Bird and bat feet and also bat wings, particularly their thumb claws, become entangled in this type of netting. But even animals without claws such as brown snakes can also become hopelessly entangled in this open-weave netting.

Safe or wildlife-friendly fruit tree netting has a finer weave – with mesh size of less than 5 mm. To determine if netting is wildlife friendly or not, simply do the netting 'finger test':

- the wrong kind of netting – you can poke your finger through the weave
- the right kind of netting – you cannot poke your finger through the weave.

The use of wildlife-friendly netting is enhanced by using a pole or frame to ensure that the netting is taut, creating a netting cage or aviary for the plant. Fruit Saver® nets, which are suitable for small and medium trees, come with an inbuilt frame. Other wildlife friendly nets include netting that is designed for protection from hail, and VegeNet®.

Here are some other tips:

- White is the preferred colour as it is more visible at night.
- Remember: these nets are plastic, and once discarded do not break down, so please dispose of them safely.

Discarded fishing tackle and other rubbish

Most of us are aware of the menacing mountain of plastic we as an industrialised society have created since the 1950s. One study estimates that we have created a staggering 6300 million tonnes of plastic, with just 9% recycled and 12% incinerated, leaving a massive 79% of plastic in landfill or the natural environment.[5]

The effects of discarded plastic like plastic bags, ghost nets and fishing tackle are well documented for marine wildlife such as dolphins, seabirds and sea turtles, but

Table 12. Threats from discarded material and solutions.

Type of material	Wildlife most affected	Solutions
Fishing line, hooks	Green sea turtles, pelicans, gulls, black swans, coastal birds of prey	Never drop fishing line on the ground. Instead, roll it up into a tight ball and tie it off, or cut it into short lengths before dropping into a rubbish bin, as this prevents birds that are scavenging at the rubbish dump from becoming entangled. Alternatively, incinerate the line by using a lighter.[10]
Human hair	Pigeons, gulls, magpie-larks and plovers	Somewhat surprisingly, bird legs get entangled in human hair the same way as fine filament fishing line! It is best to cut it in shorter lengths and dispose of it in the bin.
Plastic rings from jars, hair ties and elastic bands, disposable face masks	Platypuses, green sea turtles, cormorants, ducks	Use a pair of scissors to cut rings, hair ties and mask ear loops before placing them in the bin.

other research shows that freshwater and terrestrial species are affected by the same type of plastic encounters described for marine species, such as cormorants, ibis and monitor lizards entangled in discarded bottlecap rings.[6] Birds of many kinds, including coastal raptors such as white-bellied sea eagles, are affected by fishing line waste.[7] Other common household items include plastic bags, thread and elastic bands, and since the coronavirus pandemic – discarded face masks![8] Platypuses can become entangled in household rubber bands, hair ties and plastic rings, which cut off circulation and create deep wounds.[9]

A recent study by the Australian Platypus Conservancy found that entanglement frequency was eight times higher in metropolitan Melbourne than regional Victoria.[9] This makes sense as a higher population equals higher rates of rubbish entering the waterways, but it is also another example where urban dwellers have a great opportunity to help wildlife, simply by disposing of their rubbish thoughtfully. Similarly, residents who live directly on the foreshore, in canal-style developments such as Moreton Bay, Qld, have the opportunity to help wildlife of the greater bay area via correctly disposing of their rubbish. Table 12 lists various discarded materials and how to safely dispose of these to avoid entanglement with wildlife.

References

1. Maclean J (2011) The 'Devil's rope': flying-foxes in barbed wire fences. In *The Biology and Conservation of Australasian Bats*. (Eds B Law, P Eby, D Lunney, L Lumsden) pp. 421–423. Royal Zoological Society of New South Wales, Mosman.
2. Wildlife friendly fencing and netting, <https://wildlifefriendlyfencing.org/> (accessed 2 December 2023).
3. Scheelings TF, Frith SE (2015) Anthropogenic factors are the major cause of hospital admission of a threatened species, the grey-headed flying fox (*Pteropus poliocephalus*), in Victoria, Australia. *PLoS One* **10**(7), e0133638. doi:10.1371/journal.pone.0133638

4. van der Ree R (1999) Barbed wire fencing as a hazard for wildlife. *Victorian Naturalist* **116**(6), 210–217.
5. Geyer R, Jambeck JR, Law KL (2017) Production, use, and fate of all plastics ever made. *Science Advances* **3**(7), e1700782. doi:10.1126/sciadv.1700782
6. Blettler MCM, Mitchell C (2021) Dangerous traps: macroplastic encounters affecting freshwater and terrestrial wildlife. *The Science of the Total Environment* **798**, 149317. doi:10.1016/j.scitotenv.2021.149317
7. Thomson VK, Jones D, McBroom J, Lilleyman A, Pyne M (2020) Hospital admissions of Australian coastal raptors show fishing equipment entanglement is an important threat. *The Journal of Raptor Research* **54**(4), 414–423. doi:10.3356/0892-1016-54.4.414
8. Patrício Silva AL, Prata JC, Mouneyrac C, Barcelò D, Duarte AC, *et al.* (2021) Risks of Covid-19 face masks to wildlife: present and future research needs. *The Science of the Total Environment* **792**, 148505. doi:10.1016/j.scitotenv.2021.148505
9. Serena M, Williams GA (2022) Factors affecting the frequency and outcome of platypus entanglement by human rubbish. *Australian Mammalogy* **44**(1), 81–86. doi:10.1071/AM21004
10. Wild Bird Rescues Gold Coast. <https://wildbirdrescues.com.au/threats-to-native-birds/> (accessed 2 December 2023).

Eamon's mother was killed by a car, and Eamon was discovered stuck in the car's grill! Here he is thriving in care at Australia Zoo Wildlife Hospital.

Wildlife-friendly driving

CAR STRIKE, ROAD-KILL, WILDLIFE VEHICLE collision – the phenomenon of cars hitting animals has many names. While this book is about the wildlife-friendly house and backyard , we all travel to and from our homes, and the whole community from individuals to government can play a part in reducing the wildlife road toll.

The scale of the problem

Car strike is the leading cause of animals coming into care in Tasmania,[1] Queensland,[2] Victoria[3] and NSW,[4] and very likely in other Australian states and territories. Animals are killed directly from road trauma, or in the case of marsupials and their pouch young, orphaned when the mother is killed. The estimates are sobering reading, with some 4 million Australian mammals killed annually, and an estimated 560,000 orphaned marsupials per year.[5] Smaller wildlife such as frogs are also hit by cars in great numbers, with a staggering 40,000 frogs run over annually

PHOTO: WILDLIFE WARRIORS

on one road in northern New South Wales.[6] Car strike is also the highest cause of death for endangered species such as Tasmanian devils,[1] koalas[2] and cassowaries.[7]

All kinds of roads, from huge national highways to meandering dirt roads, also indirectly affect wildlife by forming barriers that separate populations. Many of the entries in this book describe how your backyard may make up part of an animal's home range. An echidna, koala or a python must travel through many different parts of the landscape, covering gardens, parks reserves, football fields, bushland remnants – and these areas are crisscrossed by roads.

Roads prevent wildlife from being able to use all the resources available in the landscape. Two little areas of habitat separated by a road might be too small to support a koala or brushtail possum, but if the barrier was removed, and safe movement allowed, then that population may have enough food and other resources to survive. Roads also may hamper efforts to breed. A healthy population relies on the ability of animals to disperse, as adults need to meet other individuals to breed and young animals need to venture forth and create new territories. If animal populations are too isolated from one another they risk becoming inbred and more vulnerable to disease.[8]

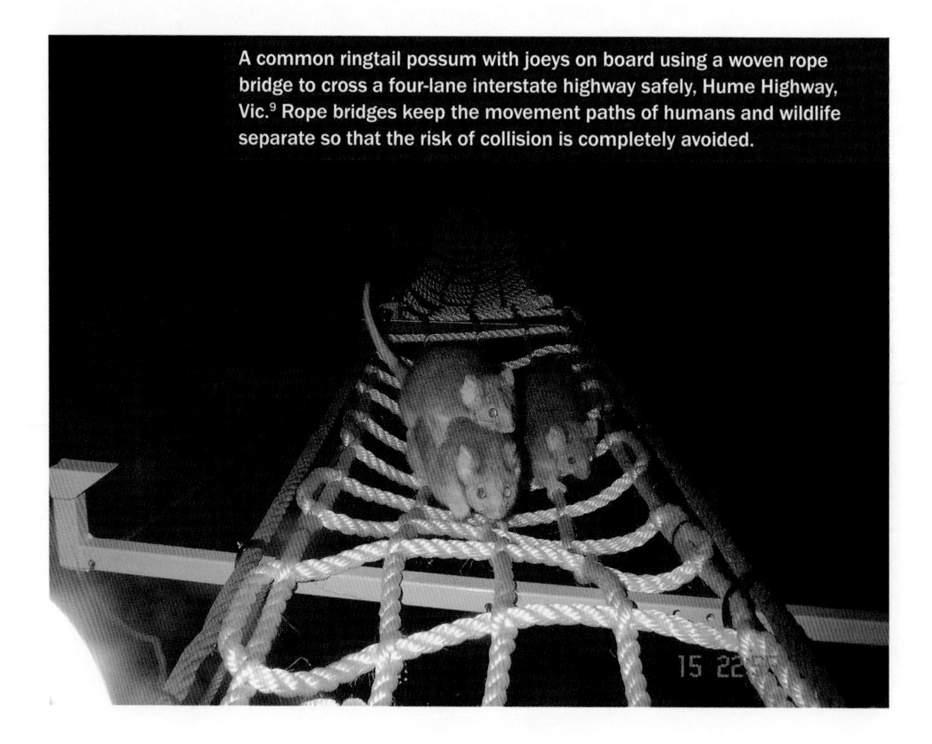

A common ringtail possum with joeys on board using a woven rope bridge to cross a four-lane interstate highway safely, Hume Highway, Vic.[9] Rope bridges keep the movement paths of humans and wildlife separate so that the risk of collision is completely avoided.

KYLIE SOANES

Actions and solutions

Reduce your driving speed

As long ago as 1960, the great naturalist David Fleay urged us to think of driving through areas with wildlife in the same way as we drive around a school zone – slowly and always scanning for the vulnerable![10]

How much slower you need to drive depends upon the animals you are trying to avoid and the context. I live in the Wombat Forest, Vic, and the 50-km winding road between my workplace and my home goes though forest patches, open farmlands and country estates. The speed limit is posted at 80–100 km an hour. At dusk and during the night, I drive at speeds between 40 and 80 km an hour depending on the conditions, but usually at ~70 km an hour. These reduced speeds allow more time to scan the sides of the road for wildlife and increased reaction times to respond to any encounters.

Driving slower does make a difference. In the early 2000s, researchers surveyed over 15,000 km of Tasmanian roads. They compared the numbers of road-killed animals and the speed limits of the road they were found on, and concluded that at a speed of 100 km an hour, a reduction of speed of just 20% (to 80 km an hour) would result in a reduction of road killed animals of around 50%![11]

A reduced speed gives you greater response time and braking distance, which are much better for driver safety, as swerving to avoid hitting an animal can often result in the driver hitting other road users or stationary objects such as a tree or power pole.[12]

If you are scanning and see animals crossing the road or about to cross, lower your high-beam headlights, as these tend to dazzle and confuse animals further. Animals such as kangaroos move in groups. If safe to do so, you can even pull over and wait for the mob to pass.

While travelling on less busy roads, if people are tailgating you impatiently as you drive slowly, pull over if it is safe to do so and let them pass. Otherwise, if you are in a wildlife area drive slowly and with pride!

When and where to slow down

How do you determine when you are driving through a wildlife area? Any patches of natural vegetation, whether forest, woodland, wetland areas or coastal scrub are likely crossing areas. Yellow wildlife road signs indicate that you are driving through an area with wildlife, and where wildlife has been killed on the road. These are often sections of the road with low visibility such as over crests, around corners or in densely vegetated patches. The presence of road-killed animals is also a clue.

At dusk and dawn, and certain times of year

Much of the wildlife we call nocturnal is actually crepuscular, which means they are most active at dawn and dusk. This includes kangaroos, wallabies and even echidnas feeding or travelling around their home ranges.

Analysis of over 17,000 AAMI animal collision claims across the country between 1 January and 31 December 2022 confirmed that dusk is the most dangerous time for wildlife related road accidents, with more than a quarter (26%) of accidents taking place between the hours of 4:30 pm and 8 pm.[13]

During the spring and summer months, many animals are active and in breeding season. This very busy period is called trauma season by many wildlife hospitals and wildlife rescue organisations. Lizards and snakes emerge in the warmer months, and young birds have left the nest and are out exploring with parents. In particular, magpies and ducks, as well as koalas, breed at this time.

In some areas, the initial spring and summer peak is followed by another peak as young animals disperse from their parent's territories. In the Tasmanian study mentioned above, road-kill occurrence actually peaked in late summer/autumn as this is when the young of the most common species (brushtail possum, Tasmanian pademelon and Bennett's wallaby) disperse.[11]

Another time to be particularly aware of wildlife on roads is during periods of extreme weather, such as heat waves, after bushfires or during floods, as animals may be displaced or travelling further afield seeking food or water (see pp. 176–182 on extreme weather).

See pp. 142–147 for what to do if you hit an animal or come across a car strike victim.

X marks the spot – orphaned wildlife

You may have seen a white, pink or yellow X spray painted on kangaroos and wombats who lay dead on the side of the road. This means that a wildlife rescuer, usually a volunteer, has stopped and checked the animal for pouch young that may still be alive. If the joey is rescued, dead or if the animal is a male, the body is marked so that other rescuers need not stop and check.

Pull dead animals off the road – if safe to do so

Road-killed animals are a source of food for a wide range of wildlife including Tasmanian devils, quolls, wedge-tailed eagles, owls, ravens, little eagles and even wallabies. Some of these animals, in turn, are killed as they feed upon the animal as they cannot get out of the way in time.

Animals also get killed when they are trying to attend to stricken members of their flock or mob, mates and their young. Birds such as cockatoos and galahs and the ever-faithful bobtail lizard may be killed while guarding or trying to protect their dead mate. Removing dead animals from the road if at all possible helps avoid these scenarios.

References

1. Heathcote G, Hobday AJ, Spaulding M, Gard M, Irons G (2019) Citizen reporting of wildlife interactions can improve impact-reduction programs and support wildlife carers. *Wildlife Research* **46**, 415–428. doi:10.1071/WR18127
2. Taylor-Brown A, Booth R, Gillett A, Mealy E, Ogbourne SM, *et al.* (2019) The impact of human activities on Australian wildlife. *PLoS One* **14**(1), e0206958. doi:10.1371/journal.pone.0206958
3. Camprasse ECM, Klapperstueck M, Cardilini APA (2023) Wildlife emergency response services data provide insights into human and non-human threats to wildlife and the response to those threats. *Diversity* **15**, 683. doi:10.3390/d15050683
4. Kwok ABC, Haering R, Travers SK, Stathis P (2021) Trends in wildlife rehabilitation rescues and animal fate across a six-year period in New South Wales, Australia. *PLoS One* **16**(9), e0257209. doi:10.1371/journal.pone.0257209
5. Englefield B, Starling M, McGreevy P (2018) A review of roadkill rescue: who cares for the mental, physical and financial welfare of Australian wildlife carers? *Wildlife Research* **45**(2), 103–118. doi:10.1071/WR17099
6. Goldingay RL, Taylor BD (2006) How many frogs are killed on a road in north-east New South Wales. *Australian Zoologist* **33**, 332–336. doi:10.7882/AZ.2006.006
7. National Environmental Science Program Threatened Species Research Hub (2019) 'Threatened Species Strategy Year 3 scorecard – southern cassowary'. Australian Government, Canberra.
8. Taylor B, Goldingay R (2010) Roads and wildlife: impacts, mitigation and implications for wildlife management in Australia. *Wildlife Research* **37**(4), 320–331. doi:10.1071/WR09171
9. Soanes K, Vesk PA, van der Ree R (2015) Monitoring the use of road-crossing structures by arboreal marsupials: insights gained from motion-triggered cameras and passive integrated transponder (PIT) tags. *Wildlife Research* **42**, 241–256. doi:10.1071/WR14067
10. Fleay D (1960) *Living with Animals*. Lansdowne Press, Melbourne.
11. Hobday AJ, Minstrell ML (2008) Distribution and abundance of roadkill on Tasmanian highways: human management options. *Wildlife Research* **35**(7), 712–726. doi:10.1071/WR08067
12. Ang JY, Gabbe B, Cameron P, Beck B (2019) Animal-vehicle collisions in Victoria, Australia: an under-recognised cause of road traffic crashes. *Emergency Medicine Australasia* **31**(5), 851–855. doi:10.1111/1742-6723.13361
13. Kahl D(2023) 'Crashing into wildlife: most Aussie drivers would dangerously swerve or slam the brakes to avoid a collision' AAMI head of Motor Claims, <https://www.suncorpgroup.com.au/news/news/AAMI-animal-collision-data>.

A coppery brushtail possum reaches for food. There is a fine line between tame and aggressive in animals that are fed by humans. Atherton Tablelands, Qld.

Is it okay to feed wildlife?

SOME OF MY WILDLIFE-LOVING FRIENDS take great pleasure in feeding their local possums, parrots and shingleback lizards. And they are not alone – perhaps a third of Australians do it! But is it okay to feed wildlife? Some say yes, if it is done correctly, but the wildlife-friendly approach is firmly in the 'do not feed' camp. The provision of fresh, clean water is a healthy alternative to feeding wildlife.

Feeding wildlife, especially birds, is a billion-dollar industry in Europe and North America. In fact, the feeding of small birds, such as chickadees and robins, on energy-rich suet pudding balls during the long, cold northern winters is actively encouraged. Here in Australia, various studies report that between one-third and even half of Australian households in the areas surveyed practise feeding wildlife.[1,2]

Feeding wildlife offers many people a chance to engage directly with nature, and people get very attached to 'their' garden wildlife, with feeding time a treasured daily ritual.[3] When fed, some species become quite bold, and people love it when

PHOTO: ETHAN MANN

a wild animal takes a piece of food from their hands and then eats it with obvious enjoyment. Overwhelmingly, feeding wildlife is carried out with the best of intentions, so while this section focuses first on some of the harm that may result we will then explore the benefits of bird baths, which allow for connection with your backyard visitors with much less risk.

The case against feeding wildlife

As feeding wildlife is enjoyed by so many households in Australia, some people, such as researcher and author Darryl Jones, say it is helpful to give advice on how to feed wildlife correctly. The basic principles include feeding the right type of food, only feeding small amounts and using strict hygiene measures to avoid the spread of disease.[2,4]

Conversely, wildlife rescue organisations and wildlife hospitals around the country implore the public to stop feeding their backyard wildlife altogether, as the potential to cause harm is too great. The social media feeds of these organisations are peppered with graphic images of the disease and malnutrition that may be caused by feeding wildlife. As well as disease, many of the species featured in this book will alter their behaviour as a consequence of being fed in gardens – often to the detriment of both humans and wildlife.

The wrong kind of food

One would think that a slice of melon or apple is perfectly fine to feed to a ringtail possum. But both common and western ringtails eat mainly leaves, flowers and gum nuts supplemented with very low sugar native fruits. Even though ringtail possums will readily eat exotic fruit, their gut biota struggles to process the high sugar content, producing gas and swelling, and sometimes even death.[5]

In other cases, food is accepted readily by the animals, and the animal subsequently suffers from nutritional deficiencies. Often when Australian magpies are offered human food, it is only a small part of their diet and they continue to forage for wild-caught prey,[6] but in some cases the parents switch to eating a lot of human food, with tragic consequences. Metabolic bone disease is a completely avoidable disease that affects baby magpies who are raised by parents that are eating a lot of human-offered mince. Unlike worms, beetles and other natural foods, mince lacks vital ingredients, in particular calcium. The chicks hatch with unformed and easily broken bones, deformed feather growth and even brittle bills that snap easily. The young magpies have to be euthanised to end their suffering. Even high-quality steak is not a complete food for magpies – they need the wholefood goodness of natural dietary items such as insects and earthworms.

Table 13. Types of food offered to wildlife, risks and alternatives.

Type of food	Species affected	Results	Advice/alternatives
Fruit	Quenda/bandicoots Bobtails/shinglebacks Ringtail possums	Dental decay Sinus infections Abscesses Obesity	Plant native fruiting species instead such as lilly pilly.
Mince and other raw meat	Australian magpies and their young	Metabolic bone disease	Offer a bird bath or buy mealworms from the pet shop.
Bird seed (especially sunflower seed)	Parrots such as galahs, gang-gang cockatoos, Australian king-parrots, lorikeets	Fatty liver Gastric yeast syndrome Spread of psittacine beak and feather disease (PBFD) around feeding stations Damage to delicate tongue and beak (lorikeets)	Offer a suite of bird baths. Replace any feeding stations with a sprinkle of seed on the lawn or balcony, or offer alternatives such as cherry tomatoes or silverbeet.
Honey/water mixes	Lorikeets Red wattlebirds	Spread of PBFD around feeding stations Nutritional deficiencies such as thiamine	Plant profusely flowering local species, as advised by your local wildlife rescue organisation or native nursery.
Bread	Kangaroos and wallabies Ducks and swans Rainbow lorikeets and other parrots	Lumpy jaw syndrome Angel wing	Bread is a processed food and should never be fed to wildlife. Some duck and swan feeders find it hard to break the habit – but can replace bread with peas and corn.

Other unsuitable but very commonly offered foods include 'bird seed' (especially sunflower seed), honey/water mixes and bread. Table 13 describes some commonly offered garden foods and their risks to wildlife, while also suggesting alternatives.

Even if the correct foods are given, feeding wildlife can have some unintended consequences.

Too much food

Like a Labrador that cannot self-regulate how much food to eat, it may come as a surprise that some wildlife also cannot regulate their food intake and will suffer from obesity if constantly offered food. This includes shinglebacks (bobtails) and bandicoots. One bobtail came into Kanyana Wildlife Rehabilitation Centre severely overweight. A bobtail's usual weight range is 250–500 g, but this lizard was 674 g. The poor fellow could hardly move due to the extra fat all over his body – even his eyelids. His claws were overgrown as he wasn't walking around to forage for his own food! The staff at Kanyana also received a very hefty quenda into care who was so overweight from eating human-offered food that he was limping from arthritis. Male quendas may weigh up to 1.6 kg, but this individual came in at more than 2.5 kg.[7]

Both animals needed time in care to reduce their weight, meaning time away from their mates and everyday lives as wild animals; this could have easily been avoided.

Disease risk

Bird seed feeding stations, particularly those that allow birds to stand in the seed, risk the spread of several diseases, including beak and feather disease (see pp. 64–68), spironucleosis, which is affecting Australian king-parrots, and psittacosis or parrot fever. Psittacosis is a zoonotic disease, meaning it can transmit to humans if we breathe in the dust from droppings and feathers of affected birds.

A fact sheet from Wildlife Health Australia gives advice on how to avoid disease transmission at bird feeders – basically they must be cleaned and disinfected as regularly as your pet's water and food dishes – daily![4] If you cannot commit to daily cleaning then it is best to avoid feeding wildlife altogether.

Increased conflict with humans

Studies have shown that when fed artificially, some brushtail possum young do not disperse into new territories and the local possum population becomes ever more

Wildlife Health Australia's advice on avoiding disease transmission

Feeding stations should be kept scrupulously clean. Stations should be thoroughly cleaned after every feeding event. Cleaning should preferably occur in an outdoor area and may require several steps, such as:

- Scrape up and disposing of remnant food and other organic material.
- Clean the feeding surface with detergent and warm water and rinsing clean. A disinfectant may be necessary; follow the directions on the container.
- Allow surfaces to fully dry before food is offered again.
- Carefully dispose of waste material in a closed bin or pit.
- Take care not to use excessive scrubbing or high water pressure when cleaning, as this might aerosolise zoonotic pathogens such as chlamydia (the cause of psittacosis).
- Always wash your hands before and after cleaning food and water containers.

abundant. This causes rising stress in the possums as they fight for limited resources such as food and shelter. High possum numbers also increase stress in those neighbours adjoining enthusiastic possum feeders who may not want an abundance of local possums. Frustrated neighbours may then respond to overabundant possums via illegal trapping and even killing of possums.[8]

The bold behaviour that is so appealing to us is not because the animal has become tame like a domestic animal, but rather because they have overcome their natural fear of humans. This means that being bold can elevate into being aggressive as the animal begins to simply demand food.[9]

This can be very frightening and potentially very dangerous when it occurs in large animals, such as cassowaries and eastern grey kangaroos. Cassowaries who have been fed may become aggressive and attack humans when in search of food (see pp. 130–135). Kangaroos can also become demanding and aggressive (see pp. 89–95).

A recent paper calling for more research into the practice of wildlife feeding noted that only the boldest individuals in a population will overcome a fear of humans to accept offered food. This may be driving artificial selection of these bolder personality types, which are often associated with increased use and better adaption to urbanised areas. We don't know what this selection for the boldest of personality types will do to the overall population.[10]

Another example of increased conflict with humans as a result of feeding is in the case of cockatoo damage. When fed, sulphur-crested cockatoos may become sedentary instead of flying around their home ranges searching for food, and damage to houses often occurs in an area where someone is regularly feeding cockatoos (see pp. 38–42).

A positive alternative: bird baths

Bird baths allow us to observe wildlife at close quarters, carrying out natural behaviours such as drinking, bathing and preening on bushes nearby afterwards. Bird baths are important any time of year, even in the cold winter months of southern Australia. And they are absolutely essential during periods of drought and during heat waves. The provision of cool, clean water during extreme heat periods and after bushfires literally saves lives (see pp. 176–182 on extreme weather).

Happily, a small-scale study carried out in Melbourne gardens revealed that if you take the leap and transition from feeding wildlife to simply providing water, you can expect to see the same suite of your favourite bird species visiting your bird baths, albeit in reduced numbers.[11] Providing water over food may benefit some of the smaller birds. In the study, thornbills and eastern spinebills were more frequently recorded at bird baths than at feeding stations.

And bird baths benefit more than birds! 'Bird baths' is actually a misnomer, as they can provide essential drinking and bathing water for many of your garden visitors – from the smallest insects to larger animals, such as wallabies. The trick is to offer a wide variety of water receptacles as different species have different preferences. All bird baths need a rock or stick inside to allow animals to climb out again.

Managing a suite of bird baths involves keeping them scrupulously clean by scrubbing with a stiff brush occasionally and replacing the water regularly. Refilling the bird bath is very rewarding, as sometimes the animals are waiting nearby for their fresh water, or even forming an orderly queue!

Type of bird bath

Shallow bird baths placed at ground level will provide water for small birds, such as fairy-wrens, and ground living fauna, such as echidnas, bandicoots and blue-tongue lizards. If you suspect your garden has wandering cats, avoid ground-level baths.

Pedestal bird baths are a popular choice and commercially available. The provision of a stick or some rocks is important as many are glazed ceramic and can be very slippery. Avoid bird baths made of metal as the water may get too hot in summer.

A Tasmanian native-hen enjoys a bird bath placed at ground level.

HELEN GREENWOOD

Placement

All bird baths need to be located near trees or shrubs, which provide shelter and perching spaces for small birds. Small birds need to feel safe when using a bird bath, and shrubs provide a few vantage points to allow the birds to check for danger before drinking or entering the water. A pedestal bird bath located in the centre of the garden in an open grassy area will be visited by only the boldest of birds and other animals.

To prevent birds flying into windows after their bath, place baths either very close to the window (0.5 m) or well away (see pp. 16–21).

Bird baths placed near the house allow for plenty of enjoyment as you watch through the windows, but consider a deeper bird bath away from the house in a quieter part of the garden for your more timid backyard visitors.

As well as a bird bath, a diverse native garden with plenty of natural food is also a great alternative to feeding wildlife. Contact your local Gardens for Wildlife initiative or your local indigenous plant nursery as they will be able to advise on the best species to plant for your local wildlife.

Bird baths: not just for the birds. Other backyard visitors such as this red-necked wallaby will drink from bird baths if they feel safe.

PIXABAY/SANDID

References

1. Jones D (2011) An appetite for connection: why we need to understand the effect and value of feeding wild birds. *Emu – Austral Ornithology* **111**(2), i–vii. doi:10.1071/MUv111n2_ED
2. Jones D (2018) *The Birds at My Table: Why We Feed Wild Birds and Why It Matters*. Cornell University Press, NewSouth Publishing, Sydney.
3. Brock M, Perino G, Sugden R (2017) The warden attitude: an investigation of the value of interaction with everyday wildlife. *Environmental and Resource Economics* **67**(1), 127–155. doi:10.1007/s10640-015-9979-9
4. Wildlife Health Australia (2020) 'Biosecurity concerns with feeding wild birds in Australia'. Canberra, <https://wildlifehealthaustralia.com.au/Portals/0/ResourceCentre/FactSheets/Avian/Biosecurity_concerns_associated_with_feeding_wild_birds.pdf>.
5. Department of Planning, Industry and Environment (2021) *Initial Treatment and Care Guidelines for Rescued Possums and Gliders*, <https://www.environment.nsw.gov.au/-/media/OEH/Corporate-Site/Documents/Animals-and-plants/Native-animals/rescued-possums-gliders-treatment-care-guidelines-210312.pdf>.
6. O'Leary R, Jones DN (2006) Use of supplementary foods by Australian magpies *Gymnorhina tibicen*: implications for wildlife feeding in suburban environments. *Austral Ecology* **31**(2), 208–216. doi:10.1111/j.1442-9993.2006.01583.x
7. Jackson C, Bobtail Coordinator, Kanyana Wildlife Rehabilitation Centre, Perth, personal communication, 20 September 2023.
8. Wilks S, Russell T, Eymann J (2008) Valued guest or vilified pest? How attitudes towards urban brushtail possums *Trichosurus vulpecula* fit into general perceptions of animals. *Australian Zoologist* **34**, 33–44.
9. Temby I (2005) *Wild Neighbours: The Humane Approach to Living with Wildlife*. Citrus Press, Sydney.
10. Griffin LL, Ciuti S (2023) Should we feed wildlife? A call for further research into this recreational activity. *Conservation Science and Practice* **5**(7), e12958. doi:10.1111/csp2.12958
11. Miller KK, Blaszczynski VN, Weston MA (2015) Feeding wild birds in gardens: a test of water versus food. *Ecological Management & Restoration* **16**(2), 156–158. doi:10.1111/emr.12157

A pair of canoodling gang-gang cockatoos. These lovely parrots are under stress from increasing frequency of heat waves and bushfires.

Extreme weather

DURING THE HEAT WAVES AND bushfires of Black Summer 2019–20, survivors such as gang-gang cockatoos and superb lyrebirds found refuge in some backyard gardens. During the 2021–22 floods, pythons, possums and wallabies clung to human made structures to escape rising floodwaters. This entry provides tips on how to make sure your garden is both safe space and refuge for our backyard wildlife during times of crisis.

PLEASE STAY SAFE AND LOOK after yourself during extreme weather events. Leave animal rescue away from your home to trained rescuers. Avoid the firegrounds and never enter floodwaters.

Prepare in 'peace time'

Extreme weather events are increasing in number and severity across Australia.[1] These include heat waves, bushfires, wind and rain events (storms, months of rain) and flooding. Extreme weather events place enormous pressure on both wildlife and the wildlife rehabilitators and rescue organisations caring for affected animals.[2]

Many wildlife hospital staff and rehabilitators refer to the times outside of natural disasters as 'peace time' as heat waves, bushfires, storms and floods often produce such a volume of animals requiring care the only suitable analogy is a war zone. Peace time is the time to follow, subscribe or become a member of your local wildlife rescue organisation, as they will provide advice on how to respond during extreme weather events – what to do, and what not to do.

Peace time is also an opportunity to implement some of the measures described in this book that create a wildlife friendly backyard. Extreme weather events often cause wildlife to disperse into unfamiliar habitats, which may include your garden, and even your house. Animals may be hungry, thirsty and stressed. Displaced and stressed wildlife may be on the ground and therefore very vulnerable to attack from pets such as cats and dogs. Wildlife-friendly pet ownership is vital during heat waves, floods and after bushfires (see pp. 148–156).

Using wildlife-friendly fencing and fruit netting will also help animals from further harm during extreme weather events. Safe fencing allows for dispersal of animals and wildlife-friendly netting ensures that starving animals seeking food do not get entangled (see pp. 157–162).

Heat waves

Heat waves have been part of the Australian environment for a very long time indeed and so our wildlife has developed many ways of reducing heat stress. Ever seen birds walking around with their bills open on a hot day? Birds do not sweat, and this is their way of reducing heat in the body, as warm air leaves their open mouth. Kangaroos and wallabies will lick their paws and forearms to promote evaporative cooling. While hanging in the camp on a hot day, flying-foxes flap their wings like fans to cool themselves. Other animals alter their behaviour to help reduce their body temperature. Lizards, pythons and ringtail possums seek shady areas, and koalas uncurl their bodies to allow heat to escape from the white fur on their underside. Koalas also lay on smooth eucalypt tree trunks which may be cooler than the air, or even come to the ground to rest under shady trees. All wildlife benefit from access to cool, clear water during heat waves.

Unfortunately, these mechanisms and behaviours only work to a point. Several days in a row of over 40°C is simply too much for many animals and the physiological symptoms of heat stress intensify until death occurs.[3] Flying-foxes are particularly vulnerable to heat waves. Researchers estimated flying-fox mortality during the Black Summer heat waves and fires, and the figure is astounding; an estimated 72,175 adult and young flying-foxes died of heat stress during this period.[4] And while heat related die-offs have occurred historically, they are increasing in frequency, intensity and even range. A heat wave in 2018 in Cairns saw Tolga Bat Hospital on the Atherton

Tablelands taking in over 500 spectacled flying-fox orphans after 23,000 spectacled flying foxes died in the first-ever recorded heat wave die-off for this species.[5]

The die-offs require a huge response from land managers and wildlife rescue organisations, so researchers developed the 'Flying-fox Heat Stress Forecaster' to assist these organisations plan for and mobilise in response to extreme heat events.[6]

Die-offs in flying-foxes are noticeable as they gather in camps during the day – for species such as small birds, ringtail possums and koalas which are dispersed throughout the landscape, it is more difficult to assess the deadly impact of heat waves. The maximum temperatures that cause hyperthermia and then death are slightly different across species, but researchers suggest that the flying-fox heat stress forecaster could be used to predict heat caused mortality events for other wildlife.[6]

This is all very grim, but the good news is that by providing shade, water and safe refuge for wildlife in your backyard during heat waves you are saving lives!

Tips

Water: Offer a variety of bird baths and shallow dishes – all with rocks and sticks. Never pour water into the mouths of heat-stressed wildlife such as koalas. Flat dishes and bowls allow animals to lap. You can water the tree they are in, and the koala will drink the water as it drips down the trunk, and animals such as possums and birds will drink from the wet leaves.

For small birds, water provided in shade can be lifesaving as during heat events they have to choose between staying in the shade to reduce heat stress and entering full sun for water to avoid dehydration. A shallow bird bath in the shade is perfect.[7]

Shade: A well-watered garden with plenty of shrubs and canopy trees (if space allows) creates vital shade and cool spaces. To create even more shade, erect an umbrella or even a tarp or shade cloth.

Prevent drowning: Thirsty animals may drown trying to reach water – ensure any buckets of water, troughs or ponds that may be around have sticks or rocks to allow the animals to clamber out. If you own a pool, a wildlife escape ramp such as a heavy duty rope secured to something heavy outside the pool is a must during drought and heat waves. Make sure you regularly check your skimmer box for any wildlife who may become trapped. Any downpipes leading to your water tank need to be covered in mesh to prevent animals getting stuck in the pipe trying to reach the water or drowning in the tank.

When to call wildlife rescue: Heat stress manifests differently in different animal groups, but the main signs that may indicate a need for veterinary assistance are disorientation, staggering around instead of walking normally, drooling or loss of

A female koala drinks water by lapping at a tree trunk.

JANINE DUFFY, KOALA CLANCY FOUNDATION

consciousness. If an animal sheltering at your home has these symptoms, call a wildlife rescue organisation for advice straight away.

After a bushfire

Periods of intense heat and drought are often followed by bushfires. Historically bushfires were named for the day on which they occurred, for example Ash Wednesday or Black Saturday. The 2019–20 bushfires were so catastrophic and of such duration that they were dubbed the Black Summer bushfires. Almost 8 million ha were burnt across the country, often near urban centres. Worryingly, for the first time, refuge areas like ferny gullies and rainforest habitat also burnt. Some 54% of World Heritage listed Gondwana rainforests were in flames over the course of Black Summer – including the much-loved Binna Burra Rainforest Lodge, Qld.[8]

In January 2020, an expert panel of scientists formed to provide scientific advice to guide the federal government's response to the ecological devastation wrought by the fires. Part of their role was to create lists of priority animals (vertebrates), invertebrates, plants and ecological communities to guide the recovery effort.

A species' inclusion on the list was prompted by the extent of their habitat burnt coupled with high direct mortality as a result of the fires.[9] For example, researchers

Table 14. Priority species for management intervention after the Black Summer bushfires that may occur in our backyards, and ways to keep them safe.

	Wildlife-friendly dog and cat ownership (see pp. 148–156)	Do not feed, bird baths instead (see pp. 168–175)	Wildlife-friendly netting and fencing (see pp. 157–162)	Wildlife-friendly driving (see pp. 163–167)	Protect chickens from predators (see pp. 96–102)
Gang-gang cockatoo, south-eastern glossy black-cockatoo					
Superb lyrebird					
Grey-headed flying-fox					
Koala					
Spotted-tail quoll					

estimated that ~30% of gang-gang cockatoo forest and woodland habitat was burnt and 10% of the population died during the fires.[10] This is devastating for a long-lived parrot with a slow breeding rate. But the gang-gang cockatoo is just one of several species that we can care for directly in our backyards. The good news is that we can play a part in bushfire recovery for some of the 119 species listed, as seen in Table 14.

The simple measures outlined in this book are even more vital for these listed species, who can act as bioindicators. If these well-known and well-studied species are under stress from large-scale bushfires, then so-called 'common' wildlife such as wombats, possums, fairy-wrens and tawny frogmouths are likely to have also been affected.

What to do in the lead-up and during a bushfire is beyond the scope of this book, but by following the advice above for heat waves you can help wildlife in the backyard after a bushfire has passed through the area. Most important are the:

- control of pets
- provision of fresh clean water
- wildlife-friendly fencing and netting.

Heavy rainfall, storms and flooding

The effects of rainfall and flooding on wildlife are less understood than the effects of heat waves, drought and bushfire.[11]

Heavy rainfall, flooding and storms may destroy possum dreys and bird nests, and tree hollows and burrows may become inundated. These result in displaced animals desperately seeking shelter to dry out, and nocturnal animals such as possums may be seen wandering in the day. Animals as varied as tree frogs, possums, birds and kangaroos may take to your shed, verandah or deck for temporary respite from the rain or the floodwaters. You can help displaced wildlife during extreme wet weather events by following these tips:

- Give displaced animals space to recover.
- Control your pets.
- Check under your car before driving as there may be wombats or other animals taking shelter.
- Remember: do not approach flying-foxes, wombats, reptiles or other large animals such as eastern grey kangaroos – these animals are best left to a trained rescuer.

These common ringtail joeys were rescued by Wildlife Victoria volunteers during floods in October 2022.

- If a waterlogged bird or possum is found in the house, you can place a washing basket over the animal, covered in a towel, while you call for assistance.
- Do not offer food, water or use any heating device unless directed by a wildlife carer or vet.

References

1. Ripple WJ, Wolf C, Gregg JW, Rockström J, Newsome TM, *et al.* (2023) The 2023 state of the climate report: entering uncharted territory. *BioScience* **73**, 841–850. doi:10.1093/biosci/biad080
2. Wildlife Health Australia (2023) 'WHA fact sheet: impacts of climate change on Australian wildlife August 2023'. Canberra, <https://wildlifehealthaustralia.com.au/Portals/0/ResourceCentre/FactSheets/General/Impacts_of_climate_change_on_Australian_wildlife.pdf>.
3. James T (2020) Facultative hyperthermia during a heatwave delays injurious dehydration of an arboreal marsupial. *The Journal of Experimental Biology* **223**(5), 1–6.
4. Mo M, Roache M, Davies J, Hopper J, Pitty H, *et al.* (2022) Estimating flying-fox mortality associated with abandonments of pups and extreme heat events during the austral summer of 2019–20. *Pacific Conservation Biology* **28**(2), 124–139. doi:10.1071/PC21003
5. Tolga Bat Hospital. *Flying fox heat stress events in Australia*, <https://tolgabathospital.org/bats/heat-stress/>.
6. Ratnayake HU, Kearney MR, Govekar P, Karoly D, Welbergen JA (2019) Forecasting wildlife die-offs from extreme heat events. *Animal Conservation* **22**(4), 386–395. doi:10.1111/acv.12476
7. McKechnie AE, Hockey PAR, Wolf BO (2012) Feeling the heat: Australian landbirds and climate change. *Emu* **112**(2), i–vii. doi:10.1071/MUv112n2_ED
8. Ward M, Tulloch AIT, Radford JQ, Williams BA, Reside AE, *et al.* (2020) Impact of 2019–2020 mega-fires on Australian fauna habitat. *Nature Ecology & Evolution* **4**, 1321–1326. doi:10.1038/s41559-020-1251-1
9. Legge S, Woinarski J, Garnett S, Nimmo D, Scheele B, *et al.* (2020) 'Rapid analysis of impacts of the 2019–20 fires on animal species, and prioritisation of species for management response'. Report prepared for the Wildlife and Threatened Species Bushfire Recovery Expert Panel. Department of Agriculture, Water and the Environment, Canberra.
10. Cameron M, Loyn RH, Oliver D, Garnett ST (2021) Gang-gang cockatoo *Callocephalon fimbriatum*. In *The Action Plan for Australian Birds 2020*. (Eds ST Garnett and GB Baker) pp. 410–413. CSIRO Publishing, Melbourne.
11. Ritchie E, Jolly C (2022) 'The sad reality is many don't survive': how floods affect wildlife, and how you can help them. *The Conversation*. March 2022 (online).

Conclusion

'Ecology is the study of the relationships between plants, animals, people, and their environment, and the balances between these relationships' – Macquarie *Dictionary*

The science of ecology is so well placed to help us understand our backyard wildlife. Throughout this book, the advice given has been drawn from scientific research or, more specifically, ecological studies that provide insight into how animals respond to urbanisation and the impact that human activities may have on their behaviour and survival. The impact of cats on wildlife, the fate of relocated snakes and possums, the amazing love lives of lizards such as shinglebacks – these studies

A shared landscape. The coastal town of Caloundra, Qld.

PHOTO: ISTOCK.COM/M VALIGURSKY

have been carried out painstakingly by researchers, sometimes over decades. Ecology is a vital tool which helps us to make informed decisions about sharing our lives with wildlife. But behaviour change requires more than simply correct information!

Living in harmony with wildlife requires focus on our values and ideas, and the support mechanisms we need to make change: in short, the human dimensions of wildlife management.

Gardens for Wildlife

One program that does just that is the Gardens for Wildlife model. Usually delivered by local government Gardens for Wildlife programs use an innovative community-based approach to support people on their journey to become habitat gardeners, creating wildlife-friendly gardens by planting local native species, removing weeds, retaining trees and mature vegetation, planting in layers from groundcover to canopy and adding habitat elements such as bird baths and nest boxes.[1]

Participants are supported by practical initiatives with a range of actions including:

- social and media promotion
- vouchers for native plants from local nursery
- education on wildlife gardening for participants
- volunteer visit for site assessment
- wildlife gardening signs to showcase participation
- citizen science programs focused on iconic local species such as bandicoots.[2]

Evaluation of the programs have revealed benefits not just for local plants and wildlife, but tangible benefits for humans too including increased wellbeing.[1]

The main focus of Gardens for Wildlife is, of course, the gardening aspect – what to plant, habitat features, with an add-on, bullet-pointed list of how to keep wildlife safe. Keeping cats indoors, dogs under control and avoiding pesticides and snail bait are usually mentioned. It is hoped that this book helps Gardens for Wildlife educators and practitioners with practical tools on how to reduce harm such as avoiding rodenticide, reducing window collision, and resolving potential conflicts such as possums in roofs.

Gardens for Wildlife programs are constrained by the resources of the local government in question. Unfortunately the greatest need for wildlife education and support is in the less-well-resourced councils with rapidly growing populations and a rapidly expanding wildlife–human interface as development spreads. These programs also put the onus on the individual, the private landholder – what about *all* our living spaces?

Breaking down the divide between city and rural and natural and non-natural

We have seen in this book that animals such as pythons do not care one bit that aviary birds are not a 'natural' source of food and that critically endangered western ringtails are just as likely to use exotic vegetation as indigenous plantings for nesting and foraging. Animals do not make the distinction between natural and artificial and this means that we have a wonderful opportunity to adopt a whole-hearted change of approach that breaks down the artificial divide between cities and 'nature', urban and suburban, by incorporating wildlife considerations into the very fabric of our houses, streets and neighbourhoods.

Part of this approach is called Biodiversity Sensitive Urban Design (BSUD), which treats biodiversity (including wildlife) as an opportunity and valued resource to be preserved and maximised at all stages of planning and design. This takes the onus from the individual landholder and places wildlife-friendly initiatives that promote coexistence in the hands of those that shape our cities.
BSUD has five principles:

1. protect and create habitat
2. help species disperse
3. minimise anthropogenic threats
4. promote ecological processes
5. encourage positive human-nature interactions.[3]

In the introduction I talked about the bulldozing of my favourite birdwatching spot - a river mouth lined with mangroves. Biodiversity sensitive urban design principles might seem contrary to such canal-style developments, where the mangrove is removed entirely and the earth terraformed to create a pad for a house and garden. But BSUD principles *can* be relevant for the new owners even here - perhaps the local council has wildlife friendly fencing by-laws that help koalas disperse, or a 24-hour cat curfew. Or perhaps the streets of the development were designed to curve around remnant large old eucalypts and speed bumps installed to slow traffic and reduce the risk of car strike.

How do we make this our future?

We have to demand it. All levels of government have a responsibility to care for wildlife and habitat, with strong nature laws and the delivery of programs that foster coexistence between wildlife and our rapidly urbanising population. Your local council is a very good place to start. The level of engagement each council has with biodiversity and wildlife issues depends partly on resources but also on community expectations which lead into council strategic plans. When your council carries out community consultation for strategic plans and precinct planning, get involved as wildlife and biodiversity-centric input is invaluable

The opinions and whims of local councillors can also exert great influence; for example, a community response to flying-fox camps and influxes is often closely related to a combination of councillor opinion and the local media.

Write to your local councillors and ask for education programs that raise awareness and provide information about coexistence with wildlife and monetary support for people wishing to make the right changes. Practical support may include subsidies to build cat enclosures or catios, or wildlife-friendly netting swaps.

Caring for our wild neighbours

I always enjoy hearing people's stories of the animals they share their lives with, and their deep concern if anything goes wrong. So deep is the connection that many people name their local wildlife! I have heard stories of Patrice and Albert the masked lapwings, Fluffy the python, Barry the catbird, Brian the green tree frog, and Poundcake the southern cassowary.

The naming of a pair of magpies that raise successive generations at your home means you have empathy and care about them – like a pet dog, a neighbour or a member of the family. Naming wildlife reflects that your lives are intertwined. Once an animal is named they can be a subject of conversation.

Naming wildlife acknowledges that rather than randomly occupying an area, many animals are entirely dependent on the same area for their whole lives. Walter the swamp wallaby is in fact quite likely to be the same wallaby every day because your property is part of his territory. As we have seen in so many entries in this book, your backyard is just one part of an animal's world. Naming the animals we live with exemplifies the notion that we are in a shared landscape.

It is hoped that this book gives you a sense of the way in which our lives and the lives of Australian wildlife intersect – 'we are all in this together', as they say. Wildlife rescuers, vets and vet nurses, rehabilitators, educators and transporters are motivated by a moral obligation to help wildlife struggling with the threats faced, often at great personal cost.[4] The advice within these pages aims to engender a similar feeling of stewardship towards your backyard wildlife and, as said in the introduction – be a wildlife 'preventer', by keeping animals safe and out of the wildlife rescue and rehabilitation system.

References

1. Mumaw L, Mata L (2022) Wildlife gardening: an urban nexus of social and ecological relationships. *Frontiers in Ecology and the Environment* **20**(6), 379–385. doi:10.1002/fee.2484
2. Selinske MJ, Bekessy SA, Geary WL, Faulkner R, Hames F, *et al.* (2022) Projecting biodiversity benefits of conservation behaviour-change programs. *Conservation Biology* **36**(3), e13845. doi:10.1111/cobi.13845

3. Garrard GE, Williams NSG, Mata L, Thomas J, Bekessy SA (2018) Biodiversity sensitive urban design. *Conservation Letters* **11**(2), e12411. doi:10.1111/conl.12411
4. Englefield B, Starling M, McGreevy P (2018) A review of roadkill rescue: who cares for the mental, physical and financial welfare of Australian wildlife carers? *Wildlife Research* **45**(2), 103–118. doi:10.1071/WR17099

Further reading

Wildlife and habitat gardens

Bishop AB (2018) *Habitat: A Practical Guide to Creating a Wildlife Friendly Australian Garden*. Murdoch Books, Crows Nest.

Grant P (2003) *Habitat Garden: Attracting Wildlife to Your Garden*. ABC Books, Sydney.

Wildlife rescue and care

Freund J, Mclean J, Freund S (2023) *The Bat Hospital*. Freund Books, Atherton.

Walraven E (2023) *Care of Australian Wildlife: for Gardeners, Landholders and Wildlife Carers*. Reed New Holland, Wahroonga.

Coexistence with Australian Wildlife

Temby I (2005) *Wild Neighbours: The Humane Approach to Living with Wildlife*. Citrus Press, Sydney.

Watharow S (2013) *Living with Snakes and Other Reptiles*. CSIRO Publishing, Melbourne.

Wildlife identification and ecology

Baker A, Gynther I (Eds) (2023) *Strahan's Mammals of Australia*. 4th edn. New Holland Publishers, Wahroonga.

Cogger H (2018) *Reptiles and Amphibians of Australia*. CSIRO Publishing, Melbourne.

Davies J, Menkhorst P, Rogers D, Clarke R, Marsack P, Franklin K (2022) *The Compact Australian Bird Guide*. CSIRO Publishing, Melbourne.

Menkhorst P, Rogers D, Clarke R, Davies J, Marsack P, Franklin K (2017) *The Australian Bird Guide*. CSIRO Publishing, Melbourne.

The Urban Field Naturalist Project (2022) *A Guide to the Creatures in Your Neighbourhood*. Murdoch Books, Sydney.

Triggs B (2004) *Tracks, Scats and other Traces*. Oxford University Press Australia, Melbourne.

Index